NUREG-1792

Good Practices for Implementing Human Reliability Analysis (HRA)

Final Report

Science Applications International Corporation

Sandia National Laboratories

U.S. Nuclear Regulatory Commission
Office of Nuclear Regulatory Research
Washington, DC 20555-0001

AVAILABILITY OF REFERENCE MATERIALS
IN NRC PUBLICATIONS

Good Practices for Implementing Human Reliability Analysis

Final Report

Manuscript Completed: April 2005
Date Published: April 2005

Prepared by
A. Kolaczkowski/Science Applications International Corporation
J. Forester/Sandia National Laboratories
E. Lois/U.S. Nuclear Regulatory Commission
S. Cooper/U.S. Nuclear Regulatory Commission

Science Applications International Corporation
405 Urban Street, Suite 400
Lakewood, CO 80220

Sandia National Laboratories
P.O. Box 5800
Albuquerque, NM 87185

E. Lois, NRC Project Manager

Prepared for
Division of Risk Analysis and Applications
Office of Nuclear Regulatory Research
U.S. Nuclear Regulatory Commission
Washington, DC 20555-0001

ABSTRACT

The U.S. Nuclear Regulatory Commission is establishing "good practices" for performing human reliability analyses (HRAs) and reviewing HRAs to assess the quality of those analyses. The good practices were developed as part of the NRC's activities to address quality issues related to probabilistic risk assessment (PRA) and, as such, support the implementation of Regulatory Guide (RG) 1.200, "An Approach for Determining the Technical Adequacy of Probabilistic Risk Assessment Results for Risk-Informed Activities," dated February 2004.

The HRA good practices documented in this report are of a generic nature; that is, they are not tied to any specific methods or tools that could be employed to perform an HRA. As such, the good practices support the implementation of RG 1.200 for Level 1 and limited Level 2 internal event PRAs with the reactor at full power. Their elements are directly linked to RG 1.200, which reflects and endorses (with certain clarifications and substitutions) the "Standard for Probabilistic Risk Assessment for Nuclear Power Plant Applications" (RA-S-2002 and Addenda RA-Sa-2003) promulgated by the American Society of Mechanical Engineers, and "Probabilistic Risk Assessment (PRA) Peer Review Process Guidance" (NEI 00-02, Revision A3) promulgated by the Nuclear Energy Institute.

This report is not intended to constitute a standard and, hence, it does not provide de facto requirements; rather, this report is intended for use as a reference guide. Consequently, the authors did not write this report with the expectation that all good practices should always be met. That is, the decisions regarding which good practices are applicable — and the extent to which those practices should be met — depends on the nature of the given regulatory application. Therefore, it is important to understand that certain practices may not be applicable for a given analysis, or their applicability may be of limited scope.

FOREWORD

This report documents good practices for performing human reliability analyses (HRAs) and assessing the quality of those analyses. By "good practices," we mean those processes and individual analytical tasks and judgments that would be expected in an HRA (considering current knowledge and state-of-the-art), in order for the HRA results to sufficiently represent the anticipated operator performance as a basis for risk-informed decisions.

The U.S. Nuclear Regulatory Commission (NRC) prepared this report as part of the agency's activities to address quality issues related to probabilistic risk assessment (PRA). The work was performed by the NRC's Office of Nuclear Regulatory Research (RES) with support from Sandia National Laboratories. The good practices were developed on the basis of NRC and contractor experience in performing HRAs, reviewing HRAs (including the individual plant examinations) and improving the technology of HRA. As such, the report reflects perspectives gained from practicing HRA for more than three decades, and perspectives gained from domestic and international HRA developmental efforts (particularly during the 1990s). The report also considers the views of domestic and international stakeholders provided through interactions during the public review and comment period on the draft report.

By documenting practices that the NRC staff considers "good" for performing and assessing HRAs, this report supports the implementation of Regulatory Guide 1.200, "An Approach for Determining the Technical Adequacy of Probabilistic Risk Assessment Results for Risk-Informed Activities," dated February 2004. In this regard, the good practices provided herein should prove useful in evaluating and formulating questions about the quality of an HRA. Nonetheless, it is important to note that this report is not intended to explicitly provide questions that a reviewer should ask; rather, this report provides *the technical basis* for developing questions about the quality of an HRA.

This report is also not intended to constitute a standard and, hence, it does not provide de facto requirements; rather, this report is intended for use as *a reference guide*. Consequently, the authors did not write this report with the expectation that all good practices should always be met. That is, the decision regarding which good practices are applicable — and the extent to which those practices should be met — depends on the nature of the given regulatory application. Therefore, it is important to understand that certain practices may not be applicable for a given analysis, or their applicability may be of limited scope.

It is also important to note that the good practices are of a generic nature; that is, they are not tied to any specific methods or tools that could be employed to perform an HRA. They are written in the context of a risk assessment for commercial nuclear power plant operations occurring nominally at full power. As such, they are specifically intended for applications involving internal initiating events, but should generally also be appropriate for external initiating events. Similarly, although this report was written for full-power applications, many of the good practices will also apply to low-power and shutdown operations; however, these practices will not be sufficient for the unique characteristics of such modes of operation. In addition, elements of this report may prove beneficial in examining human actions related to nuclear materials and safeguard-related applications.

Carl J. Paperiello, Director
Office of Nuclear Regulatory Research

CONTENTS

Appendices

Tables

ACKNOWLEDGMENTS

The authors would like to thank Gareth Parry of the U.S. Nuclear Regulatory Commission (NRC) for several thorough reviews of this report at various stages of its development and for many important comments and suggestions. Similarly, we would like to thank Dennis Bley of Buttonwood Consulting, Inc.; Bruce Hallbert of Idaho National Engineering and Environmental Laboratory; and James Bongarra, Andrew Kugler, Martin Stutzke, David Lew, and Autumn Szabo of the NRC for their reviews and comments.

ABBREVIATIONS

ACRS	Advisory Committee on Reactor Safeguards (NRC)
AOP	abnormal operating procedure
ASEP	Accident Sequence Evaluation Program
ASME	American Society of Mechanical Engineers
ATHEANA	A Technique for Human Event Analysis
BWR	boiling-water reactor
CDF	core damage frequency
CESA	Commission Errors Search and Assessment
CFR	*Code of Federal Regulations*
CR	control room
CREAM	Cognitive Reliability and Error Analysis Method
DC	direct current
EOC	error of commission
EOO	error of omission
EOP	emergency operating procedure
HEP	human error probability
HERA	Human Event Repository and Analysis
HFE	human failure event
HLR	high-level requirement
HR	human reliability
HRA	human reliability analysis
HSI	human-system interface
ICDE	International Common Cause Data Exchange
IPE	Individual Plant Examination
LERF	large early release frequency
LOCA	loss-of-coolant accident
MERMOS	Methode d'Evaluation de la Realisacion des Missions Operateur la Sureté
NEI	Nuclear Energy Institute
NPP	nuclear power plant
NRC	U.S. Nuclear Regulatory Commission
PORV	power-operated relief valve
PRA	probabilistic risk assessment
PSF	performance-shaping factor
PWR	pressurized-water reactor

RES	Office of Nuclear Regulatory Research (NRC)
RG	regulatory guide
SAMG	severe accident management guideline
SCBA	self-contained breathing apparatus
SGTR	steam generator tube rupture
SLIM	Success Likelihood Index Methodology
SR	supporting requirement
SSC	system, structure, and component
TE	technical element
THERP	Technique for Human Error Rate Prediction

1. INTRODUCTION

1.1 Background

In accordance with its policy statement on the use of probabilistic risk assessment (PRA) (Ref. 1), during the past decade, the U.S. Nuclear Regulatory Commission (NRC) has increasingly been using PRA technology in "all regulatory matters to the extent supported by the state-of-the-art in PRA methods and data." Examples of risk-informed initiatives include rulemaking activities such as risk-informing Title 10, Part 50, of the *Code of Federal Regulations* (10 CFR Part 50) (Ref. 2), generating a risk-informed framework to support licensee requests for changes to a plant's licensing basis (Regulatory Guide 1.174) (Ref. 3), risk-informing the Reactor Oversight Process, performing risk studies [e.g., for steam generator tube rupture (SGTR), and fire events], and evaluating the significance of events. In addition, the NRC is using PRA in the development of an infrastructure for use in licensing new reactors.

Given the increasing importance of PRAs in regulatory decision-making, it is crucial that decision-makers have confidence in the PRA results. To address issues related to PRA quality, the NRC has issued Regulatory Guide (RG) 1.200 (Ref. 4), which describes an approach that the staff considers acceptable for use in assessing the technical adequacy of PRA results for risk-informed activities. Regulatory Guide 1.200 (Ref. 4) reflects and endorses (with certain clarifications and substitutions) guidance in standards produced by various societies and industry organizations. In particular, these include the "Standard for Probabilistic Risk Assessment for Nuclear Power Plant Applications" (RA-S-2002 and Addenda RA-Sa-2003, Ref. 5) promulgated by the American Society of Mechanical Engineers (ASME), and "Probabilistic Risk Assessment (PRA) Peer Review Process Guidance" (NEI 00-02, Revision A3, Ref. 6) promulgated by the Nuclear Energy Institute (NEI).

Regulatory Guide 1.200 (Ref. 4), the ASME Standard (Ref. 5) and NEI 00-02 (Ref. 6) provide guidance at a high level, addressing what to do, but not how to do it. Consequently, there may be several approaches to address certain analytical elements of a PRA, which meet the standards by making different assumptions and approximations and, hence, producing different results. This is particularly true of human reliability analysis (HRA), characterized by lack of consistency among practitioners on the treatment of human performance in the context of a PRA. Therefore, the guidance provided in RG 1.200 and associated documents is not sufficient to address HRA quality issues at an adequate level for regulatory decision-making. For example, the NRC's Standard Review Plan (NUREG-0800) (Ref. 7), Chapter 19, Section A.8, "Modeling of Human Performance," requires the NRC staff to determine whether "the modeling of human performance is appropriate," where "appropriate" applies to both the scope and the quality of the analysis. While RG 1.200 (Ref. 4) and associated documents may be adequate for use in determining whether the right issues are addressed, it is not adequate for use in determining how well the issues are treated. In order to address this issue, the NRC is developing HRA guidance to support the implementation of RG 1.200 (Ref. 4). However, the HRA guidance (and this report, in particular) does not address the different capability categories stipulated in RG 1.200 (Ref. 4) and associated documents. As discussed below, practitioners must determine the level of analysis needed to meet specific capability categories on a case-by-case basis.

The NRC staff is developing the HRA guidance in two phases. The first phase focuses on developing this "HRA Good Practices" report; the second phase will involve reviewing and evaluating current HRA approaches to assess their capabilities to meet the good practices when employed to address different regulatory applications.

This report provides the technical basis for evaluating the adequacy of an HRA; however, it does not specify requirements for those analyses. Depending on the application, a given analysis may not have to meet every good practice. With the good practices in mind, practitioners should determine the extent to which an analysis is adequate, although some of its elements have not completely addressed pertinent good practices. Similarly, practitioners may need to make decisions with respect to meeting the different capability categories in the ASME Standard (Ref. 5). Such judgments will require careful consideration concerning the goals of the analysis, the issues being addressed, and the importance of each good practice relative to those goals and issues. Nonetheless, knowing what the good practices are and why they are important provides a starting point for practitioners to decide what must be done, and for reviewers to determine whether the analysis was done rigorously enough to address the issue.

1.2   **Purpose**

This report provides a reference guide to good practices in HRA. By "good practices," we mean those processes and individual analytical tasks and judgments that would be expected in an HRA (considering current knowledge and state-of-the-art) in order for the HRA results to sufficiently represent the anticipated operator performance as a basis for risk-informed decisions. This report focuses primarily on the process of performing am HRA and does not, for instance, specifically address HRA data or details of specific quantification approaches, which the staff plans to address during the second phase of HRA guidance development . As such, this report has two primary uses:

(1)   Support the review of HRAs to assess the quality of analyses submitted to the NRC for decision-making. In this regard, the HRA good practices provided herein should prove useful in evaluating and formulating questions about the quality of an HRA. Nonetheless, it is important to note that this report is not intended to explicitly provide questions that a reviewer should ask; rather, this report provides *the technical basis* for developing questions (and potentially a regulatory guide and/or a standard review plan).

(2)   Provide (by implication) guidance for performing HRAs (initially or when analyzing a change to current plant practices). It supports the implementation of RG 1.200 (Ref. 4) and focuses on the attributes of a good HRA, regardless of the specific methods or tools that are used.

This document provides HRA good practices that when implemented will enable practitioners to determine the impacts of human actions as *realistically as necessary* in an assessment of risk. Note the emphasis on *necessary*, rather than *possible*. For example, depending on the purpose for which the PRA is to be used, a conservative treatment of human performance may be sufficient to address a PRA application; additional realism may not be necessary and could be a waste of resources. However, a conservative approach may *not be sufficient* when used as the basis for not needing to further investigate the issue at hand. In other words, using conservatism to justify not doing an analysis is not the same as doing enough analysis to determine that a conservative approach is adequate. Failing to adequately investigate the issue could potentially constrain the capability to identify weaknesses in plant operations and practices related to the particular human actions credited in the PRA.

This report *does not constitute a standard* and, hence, it is not intended to provide de facto requirements. Although most of the HRA good practices should be addressed to some degree in any PRA application, the authors did not write this report with the expectation that all good practices should always be met. That is, the decision concerning which good practices are applicable — and the extent to which those practices should be met — depends on the nature of the given regulatory application. Therefore, it is important to understand that certain practices may not be applicable for a given analysis, or their applicability may be of limited scope. For example, rather than assessing the degree of dependencies among human actions as described in this report, it may be more efficient and acceptable to simply assume complete dependence among the actions if the issue being addressed merely demonstrates that the resultant risk metric (e.g., core damage frequency) is no worse than some threshold value; in this case, using a conservative treatment of human action dependencies should be acceptable.

In some cases, the need to meet good practices could potentially be determined through the use of importance measures of the human actions being modeled. Documents such as NUREG-1764, "Guidance for the Review of Changes to Human Actions" (Ref. 8), provide guidance for ranking human actions by measures of importance. Such a categorization scheme could help reviewers determine the degree to which a good practice needs to be addressed. However, as previously noted, relative "importance" in the PRA should not be the only criterion for determining the needed level of analysis. Additional or other criteria may be needed because generic, non-detailed, or screening level analyses can fail to detect plant vulnerabilities and weaknesses in plant practices.

2. OVERVIEW OF GOOD PRACTICES FOR HRA

2.1 Scope of this Report

The HRA good practices are written in the context of a risk assessment for commercial nuclear power plant (NPP) operations occurring nominally at full power. The guidance is specifically for HRAs for reactor, full-power, and internal events applications, although most of the guidance should be useful for other applications (e.g., external events and other operating modes). As such, this report neither endorses nor suggests that a specific method or tool should be used, since many exist and all have strengths and limitations regarding their use and applicability. Similarly, although this report was written for full-power applications, many of the good practices will also apply to low-power and shutdown operations; however, these practices will not be sufficient for addressing the unique characteristics of such modes of operation. In addition, elements of this report may prove beneficial in examining human actions related to nuclear materials and safeguard-related applications.

The performance of an HRA typically involves several tasks or activities. Some of these tasks depend on the HRA method or quantification approach that is used. Because this report neither endorses nor specifies the use of specific HRA methods, data, or quantification approaches, most of the guidance in this document is directed at the process for performing an HRA. Nonetheless, this report does provide some non-method-specific good practices with respect to HRA quantification.

This report explicitly addresses good practices for both pre- and post-initiator event analyses, but it does not explicitly address the human contribution to initiating events. The reason is that in current NPP PRA practices, initiating event frequencies are typically developed by aggregating data from *either* equipment-induced or human-induced initiators. As a result, the contribution of human performance to the frequency of initiating events and, therefore, plant risk, is not separately analyzed in most at-power PRAs today. One exception is the analysis of support system initiating events, as discussed below. However, an explicit modeling of human-induced initiating events could potentially lead to important insights regarding the overall significance of human performance on safety and, thereby, enhance the predictability of PRA models.

As the use of PRA for decision-making continues to expand, the need to separately address human-induced initiators may emerge. In some cases, fault trees developed for the support system initiators could include human failure events (HFEs) depicting those actions that could lead to a human-induced support system failure. Such actions will have the characteristics of either pre- or post-initiating event HFEs. The techniques needed to analyze such HFEs are similar to those described herein for either pre- or post-initiating events. Therefore, this report covers the good practices needed to explicitly address human-induced initiators. [See for example, the high-level requirement (HLR) IE-C in the ASME Standard (Ref. 5), and supporting requirements such as IE-C9 concerning the modeling of recovery actions in an initiator fault tree, and IE-C12 concerning procedural influences on the interfacing system loss-of-coolant accident frequency.] However, an explicit modeling of human-induced initiating events could potentially lead to important insights with respect to the overall significance of human performance on safety and, thereby, enhance the predictability of PRA models. As the use of PRA for decision-making continues to expand, the need to separately address human-induced initiators may emerge.

2.2 HRA Good Practices and the State-of-the-Art in HRA

The HRA good practices were developed on the basis of NRC and contractor experience; lessons learned from developing HRA methods, such as the Technique for Human Error Rate Prediction (THERP) (Ref. 9), Success Likelihood Index Methodology (SLIM) (Ref. 10), and A Technique for Human Event Analysis (ATHEANA) (Ref. 11); and experience in performing and reviewing HRAs [e.g., the studies in NUREG-1150 (Ref. 12) and contractor support of several individual plant examinations (IPEs) and subsequent updates]. However, they also consider the views of domestic and international stakeholders provided through interactions during the public review and comment period on the initial draft of this report, as well as perspectives gained from domestic and international HRA developmental efforts over the years, particularly during the 1990s, such as ATHEANA (Ref. 11), Methode d'Evaluation de la Realisacion des Missions Operateur la Sureté (MERMOS) (Ref. 13), and Cognitive Reliability and Error Analysis Method (CREAM) (Ref. 14).

Thus, although RG 1.200 (Ref. 4) and the ASME Standard (Ref. 5) do not explicitly address errors of commission (EOCs), consistent with the state-of-the-art in HRA, it is recommended that future HRA/PRAs should attempt to identify and model not only errors of omission (EOOs) as is typically done, but also potentially important EOCs. Several existing sources provide useful guidance for identifying and treating EOCs in the context of PRAs, such as ATHEANA (Ref. 11), Commission Errors Search and Assessment (CESA) (Ref. 15), Julius et al. (Ref. 16), Macwan and Mosleh (Ref. 17), Vuorio and Vaurio (Ref. 18), and Wakefield (Ref. 19). This report provides guidance for identifying characteristics of situations that can facilitate EOCs, particularly in the context of risk-informed analyses being done to support plant changes. The good practices on EOCs would apply to any HRA method used for quantification.

The HRA technology continues to evolve and, as with any evolving technology, what is considered good practice today may be viewed as inferior tomorrow. Hence, although most of the good practices provided herein will always constitute good practice, some may be become outdated as the HRA state-of-the-art evolves. Therefore, this report must be considered a snapshot of HRA good practices circa 2005.

2.3 Report Organization

This report presents the good practices using a logical PRA analysis approach. As such, this report is written in a way that links the prescribed good practices to RG 1.200 (Ref. 4) and subsequent ties to requirements in the ASME Standard (Ref. 5) and particularly the HRA section of that document. However, nearly all other sections of the ASME Standard (Ref. 5) also have some parallel requirements with regard to operator actions [such as in the accident sequence analysis, success criteria, systems analysis, and large early release frequency (LERF) analysis sections]. Appendix A provides a cross-reference table that links the good practices in this report to the corresponding HRA sections in the ASME Standard (Ref. 5), RG 1.200 (Ref. 4), and NEI 00-02 (Ref. 6), as appropriate.

Chapter 3 provides overall HRA good practices. Chapters 4 and 5 provide good practices for identifying, screening, modeling, and quantifying pre- and post-initiating human events, respectively. Chapter 6 discusses good practices for modeling EOCs, while Chapter 7 focuses on documenting the analyses, and Chapter 8 provides a list of related references. Appendix A provides cross-references between the elements of this report and the corresponding elements of the ASME Standard (Ref. 5) and NEI 00-02 (Ref. 6). It also notes where RG 1.200 (Ref. 4) has added clarification or qualification to pertinent sections of the ASME Standard (Ref. 5) and NEI 00-02 (Ref. 6). Appendix B provides guidance on the consideration of performance-shaping factors (PSFs) for post-initiator HFEs. Finally, Appendix C summarizes the public comments received on the August 2004 draft of this report and highlights the NRC's responses.

2.4 Summary of HRA Good Practices

Table 2-1 below provides a summary of the good practices.

Table 2-1 Summary of HRA Good Practices

Analysis Activity	Good Practice	Section
HRA team formation and techniques for a realistic analysis	GP 1: Perform a Multi-Disciplinary, Integrated Analysis. The HRA assessment should involve a multi-disciplinary team that interacts with the rest of the PRA team in addressing each accident scenario at each stage of the analysis.	3.1.3.1
	GP 2: Perform Field Observations and Discussions. In addition to the review of plant documents, the HRA should include walkdowns of relevant actions (particularly local actions), observations of simulator exercises, talk-throughs of accident scenarios and related actions with plant operators and trainers, and other field observations and discussions, as needed.	3.1.3.2
Pre-Initiators: Identifying human actions that could leave equipment unavailable	GP 1: Review Pre-Initiator Procedures, Actions, and Equipment. All routine (scheduled) testing and maintenance, as well as calibration procedures that affect equipment to be credited in the PRA, should be identified and reviewed. Actions and equipment specified in the procedures should be examined to determine whether misalignment or miscalibration could occur and render the equipment unavailable or faulty. If so, the actions and equipment should be "considered for modeling" at least until initial screening is performed. In addition, unscheduled maintenance acts that lead to reconfiguration of a system are opportunities for restoration errors and should be addressed in the analysis. Review of historical experience (in terms of readily available databases or other sources) is also recommended as a means of identifying pre-initiators.	4.1.3.1

Analysis Activity	Good Practice	Section
	GP 2: Do Not Ignore Pre-Initiators. The identification process should identify pre-initiator human actions even if they may be potentially covered by the affected equipment failure data.	4.1.3.2
	GP 3: Examine Other Operational Modes and Routine Actions Affecting Structures (if applicable). The identification process should address other operational modes and routine actions affecting barriers and other structures such as fire doors, block walls, drains, seismic restraints, etc. (if applicable for the analysis).	4.1.3.3
	GP 4: Identify Actions Affecting Redundant and Multiple Diverse Equipment. The identification process needs to include possible pre-initiator actions *at least within each system* where redundant or multiple diverse equipment can be affected by (1) a single act, or (2) through a common failure with similar multiple acts.	4.1.3.4
Pre-Initiators: Screening human actions that do not need to be modeled	GP 1: Screen Pre-Initiators with Acceptable Restoration Mechanisms or Aids. Pre-initiators with signals, signs, or checks that help to ensure that the equipment will be reliably restored to its desired state can be screened from further analysis (with the exception noted in the next good practice below).	4.2.3.1
	GP 2: Do Not Screen Actions Affecting Redundant and Multiple Diverse Equipment. In general, do not screen those pre-initiator actions that simultaneously affect multiple (redundant or diverse) equipment items.	4.2.3.2
	GP 3: Reevaluate the Screening Process for Special Applications. For a specific PRA application such as a plant change or analysis of a special issue, revisit the original PRA to ensure that the pre-initiator screening is still valid for the current application.	4.2.3.3
Pre-Initiators: Modeling specific HFEs corresponding to the unscreened human actions	GP 1: Include HFEs for Unscreened Human Actions in the PRA Model. Define each specific pre-initiator HFE to be modeled in the PRA as a basic event that describes the human-induced failure mode and preferably locate it in the model such that it is linked closely to the unavailability of the affected component, train, system, or overall function (i.e., level of modeling) depending on the effect(s) of the HFE.	4.3.3.1
Pre-Initiators: Quantifying the corresponding human error probabilities (HEPs) for the specific HFEs	GP 1: Use Screening Values During the Initial Quantification of the HFEs. Use of screening HEPs is acceptable provided (1) it is clear that the individual values used are overestimations of the probabilities if detailed assessments were to be performed, and (2) dependencies among multiple HFEs appearing in an accident sequence are conservatively accounted for. Individual screening values should not be less than 0.01 and the joint probability of multiple HEPs in a sequence should not be lower than 0.005.	4.4.3.1

Analysis Activity	Good Practice	Section
	GP 2: Perform Detailed Assessments of Significant HFEs. To help understand the role of significant HFEs in plant safety and the factors influencing their likelihood, a detailed assessment (quantification) of at least the significant HFEs should be performed.	4.4.3.2
	GP 3: Revisit the Use of Screening Values vs. Detailed Assessments for Special Applications. For a specific PRA application such as a plant change or analysis of a special issue, revisit the original PRA to ensure that the appropriate HFEs receive detailed assessments in the new analysis.	4.4.3.3
	GP 4: Account for Plant- and Activity-Specific PSFs in the Detailed Assessments. Pre-initiator HEP assessments should account for the most relevant plant- and activity-specific PSFs in the analysis. Potentially important PSFs include written work plans, procedures, training, complexity and number of steps, reliance on memory, ergonomics, and the task environment.	4.4.3.4
	GP 5: Apply Plant-Specific Recovery Factors. To the extent the plant has "built-in" specific practices to recover any failures of pre-initiator human actions, they should be applied to the HEP evaluations for the HFEs. Multiple recoveries may be acceptable, but any dependencies among the initial failure and the recoveries, and among the recoveries themselves, must be considered. Example recovery factors include testing, independent verification, scheduled checks, and compelling signals.	4.4.3.5
	GP 6: Account for Dependencies Among the HEPs in an Accident Sequence. Dependencies should be quantitatively accounted for by deriving HEPs so that they reflect commonalities and relationships among the HFEs. A main concern is human-related "common-cause factors" that would lead to similar errors occurring across similar systems. The impact of recovery factors should be included in evaluating dependencies.	4.4.3.6
	GP 7: Assess the Uncertainty in HEPs. Point estimates should be mean values for each HEP (excluding screening HEPs) and an assessment of the uncertainty in the HEPs should be performed at least for the significant HEPs. Typical assessments of uncertainty involve developing uncertainty distributions for the HEPs, propagating uncertainty distributions through the quantitative analysis of the entire PRA, performing sensitivity analyses that demonstrate the effects on the risk results for extreme estimates in the HEPs based on at least the expected uncertainty range, or addressing through qualitative arguments. Aleatory and epistemic uncertainties should be addressed as necessary.	4.4.3.7

Analysis Activity	Good Practice	Section
	GP 8: Evaluate the Reasonableness of the HEPs Obtained Using Detailed Assessments. The pre-initiator HEPs (excluding the screening HEPs) should be reasonable from two standpoints: (1) first and foremost, relative to each other (i.e., the probabilistic ranking of the failures when compared one to another), and (2) in absolute terms (i.e., each HEP value), given the relative strengths of the positive and negative PSFs identified as being important and the presence or absence of recovery factors. Example evaluation techniques include consideration of actual plant history, comparisons with results of other analyses, and qualitative understanding of the actions and their contexts by experts.	4.4.3.8
Post-Initiators: Identifying post-initiator human actions Note: The three GPs associated with this activity are to be performed in an iterative manner and a stringent order is not implied.	GP 1: Review Post-Initiator Related Procedures and Training Materials. Plant-specific emergency operating procedures (EOPs), abnormal operating procedures (AOPs), annunciator procedures, system operating procedures, and severe accident management guidelines (SAMGs) should be reviewed. Other relevant special procedures (e.g., fire emergency procedures) should also be reviewed as appropriate. Observations of simulator exercises and talk-throughs of accident scenarios and related actions with plant operators and trainers can support the identification of post-initiator human actions at this stage. To the extent plant-specific or general industry experience is useful for identifying potential post-initiator actions of interest, it is recommended that this too be examined to provide a comprehensive set of post-initiator actions.	5.1.3.1
	GP 2: Review Functions and Associated Systems and Equipment to be Modeled in the PRA. The PRA team's plant and system knowledge should be used to identify critical functions and equipment needed and not needed for the given accident scenario. In addition, ways the equipment can functionally succeed and fail should be determined, along with (1) ways the operators are intended/required to interact with the equipment, and (2) how they are to respond to equipment failure modes that can cause undesired conditions per the PRA. During the identification process, it is helpful to use action words such as actuate, initiate, isolate, terminate, control, change, etc. so that the desired actions are clear.	5.1.3.2
	GP 3: Look for Certain Expected Types of Actions. The types of actions expected to be identified as post-initiator human actions include necessary and desired actions to directly provide a critical function, backup actions to failed automatic responses, and anticipated procedure-guided or skill-of-the-craft recovery actions. Although the actions modeled will generally be error of omission, identification of errors of commission may sometimes be important.	5.1.3.3

Analysis Activity	Good Practice	Section
Post-Initiators: Modeling specific HFEs corresponding to the human actions	GP 1: Include HFEs for Needed Human Actions in the PRA Model. Define each specific post-initiator HFE to be modeled in the PRA as a basic event that describes the human-induced failure mode and preferably locate it in the model such that it is linked closely to the unavailability of the affected component, train, system, or overall function (i.e., level of modeling) depending on the effect(s) of the HFE. The nature of the action, the consequences if its failure, the nature of the subtasks involved, and the level of detail already in the model should be considered in deciding how to model the HFEs (e.g., where it is placed and whether it should be broken into separate HFEs). Dependencies must also be considered.	5.2.3.1
	GP 2: Define the HFEs Such that they are Plant- and Accident Sequence-Specific. Each of the modeled post-initiator HFEs should be defined such that they are plant- and accident sequence-specific, and the basic events representing them are labeled uniquely. In order for the action to occur, the operator must diagnose the need to take the action and then execute the action. While many PSFs are used to quantify the probability for failing to diagnose and perform the action correctly (as discussed later under quantification), all of which should be evaluated based on plant- and accident sequence-specifics, consideration of plant- and accident-specific timing information, along with procedural and training information is critical.	5.2.3.2
	GP 3: Perform Talk-Throughs, Walkdowns, Field Observations, and Simulator Exercises (as necessary) to Support the Modeling of Specific HFEs. To fully understand the nature of the act(s) (e.g., who performs it, what is done, how long does it take, whether there are special tools needed, whether there are environmental issues or special physical needs, whether there is a preferred order of use of systems to perform a specific function, etc.) and help define the HFEs and their context, additional reviews, talk-throughs, walkdowns, field observations, and simulator exercises are performed (as discussed in Good Practice #2 under Section 3.1.3.2, with more about the benefits of these techniques presented in Appendix B). In addition, the results of these activities may add to the list of actions and/or help interpret how procedural actions should be defined based on how they are actually carried out.	5.2.3.3

Analysis Activity	Good Practice	Section
Post-Initiators: Quantifying the corresponding HEPs for the specific HFEs	GP 1: Address Both Diagnosis and Response Execution Failures. Whether using conservative or detailed estimations of the post-initiator HEPs, the evaluation should address both diagnosis and execution failures.	5.3.3.1
	GP 2: Use Screening Values During the Initial Quantification of the Post-Initiator HFEs. The use of conservative HEP estimates to screen unimportant HFEs is acceptable provided (1) it is clear that the individual values used are overestimations of the probabilities if detailed assessments were to be performed **and** (2) dependencies among multiple HFEs appearing in an accident sequence are conservatively accounted for. Individual screening values should never be less than 0.1 and the joint probability of multiple HEPs in a sequence should not be lower than 0.05.	5.3.3.2
	GP 3: Perform Detailed Assessments of Significant Post-Initiator HFEs. To help understand the role of significant HFEs in plant safety and the factors influencing their likelihood, a detailed assessment (quantification) of at least the significant HFEs should be performed. Insignificant HFEs should also be assessed in detail if detection of **all** weaknesses in plant design or practices is desirable given the application.	5.3.3.3
	GP 4: Revisit the Use of Post-Initiator Screening Values vs. Detailed Assessments for Special Applications. For a specific PRA application such as a plant change or analysis of a special issue, revisit the original PRA to ensure that the appropriate HFEs receive detailed assessments in the new analysis.	5.3.3.4
	GP 5: Account for Plant- and Activity-Specific PSFs in the Detailed Assessments of Post-Initiator HEPs. Post-initiator HEP assessments should account for the most relevant plant- and activity-specific PSFs. Potentially important main control room PSFs include (but are not limited to) procedures (and how the procedures are implemented), training, task complexity, workload, team dynamics, and scenario timing. Potentially important PSFs for local actions include (but are not limited to) procedures (and how they are implemented), training, task complexity, workload (staff available for the action), team dynamics, scenario timing, communication requirements, number of steps, reliance on memory, ergonomics, task environment, accessability, special fitness needs, and the need and location of special tools.	5.3.3.5

Analysis Activity	Good Practice	Section
	GP 6: Account for Dependencies Among Post-Initiator HFEs. Dependencies among the post-initiator HEPs in an accident sequence should be quantitatively accounted for in the PRA model by virtue of the joint probability used for the HEPs. In analyzing for possible dependencies among the HFEs in an accident sequence, look for links among the actions, including the same crew member(s) is responsible for the acts, the actions take place relatively close in time in the sense that a crew "mindset" or interpretation of the situation might carryover from one event to the next, the procedures and cues used along with the plant conditions related to performing the actions are identical (or nearly so) or related, and the applicable steps in the procedures have few or no other steps in between the applicable steps, there are similar PSFs for performing the acts, how the actions are performed is similar and they are performed in or near the same location, and the interpretation of the need for one action might bear on the crew's decision regarding the need for another action. Onceall the relationships are considered and the dependencies are included in the HEP values, the total combined probability of all the HFEs in the same accident sequence/cut set should not be below the range of ~0.0001 to 0.00001.	5.3.3.6
	GP 7: Assess the Uncertainty in HEPs. Mean values for each HEP (excluding conservative HEPs) and an assessment of the uncertainty in the HEPs are performed at least for the significant HEPs to the extent that these uncertainties need to be understood and addressed in order to make appropriate risk-related decisions. Typical assessments of uncertainty involve developing uncertainty distributions for the HEPs, propagating uncertainty distributions for the HEPs through the quantitative analysis of the entire PRA, performing sensitivity analyses that demonstrate the effects on the risk results for extreme estimates in the HEPs based on at least the expected uncertainty range, or addressing through qualitative arguments. Aleatory and epistemic uncertainties should be addressed.	5.3.3.7
	GP 8: Evaluate the Reasonableness of the HEPs Obtained Using Detailed Assessments. The post-initiator HEPs (excluding the screening HEPs) should be reasonable from two standpoints: (1) first and foremost, relative to each other (i.e., the probabilistic ranking of the failures when compared one to another), and (2) in absolute terms (i.e., each HEP value), given the context and combination of positive and negative PSFs and their relative strengths. Example evaluation techniques include consideration of actual plant history, comparisons with results of other analyses, and qualitative understanding of the actions and their contexts by experts.	5.3.3.8

Analysis Activity	Good Practice	Section
Post-Initiators: Adding recovery actions to the PRA	GP 1: Define Appropriate Recovery Actions. Based on the failed functions, systems, or components, identify recovery actions to be credited that are not already included in the PRA (e.g., aligning another backup system not already accounted for...) and that are appropriate to be tried by the crew to restore the failure. Aspects to consider include the following examples: • whether the cues will be clear and provided in time to indicate the need for a recovery action • the most logical recovery actions for the failure based on the cues that will be provided • whether the recovery is a repair action (e.g., the replacement of a motor on a valve so that it can be operated) • whether sufficient time is available • whether sufficient crew resources exist to perform the recovery(ies), • whether there is procedure guidance to perform the recovery(ies) • whether the crew has trained on the recovery action(s) including the quality and frequency of the training • whether the equipment needed to perform the recovery(ies) is accessible and in a non-threatening environment (e.g., extreme radiation) • whether the equipment needed to perform the recovery(ies) is available in the context of other failures and the initiator for the sequence/cut set	5.4.3.1
	GP 2: Account for Dependencies. All the good practices provided above for post-initiator HFEs apply specifically to recovery actions as well. Particular attention should be paid to accounting for dependencies among the HFEs including the credited recovery actions.	5.4.3.2
	GP 3: Quantify the Probability of Failing to Perform the Recovery(ies). Per the quantification good practices above, quantify the probability of failing to perform the recovery(ies) by (1) using representative data that exists and deemed appropriate for the recovery event, or (2) using the HRA method/tool(s) used for the other HFEs (i.e., using an analytical/modeling approach). If using data, ensure the data are applicable for the plant/sequence context or that the data are modified accordingly.	5.4.3.3

Analysis Activity	Good Practice	Section
Errors of Commission (EOCs)	GP 1: Address EOCs in Future HRAs/PRAs (Recommendation). Given the recent advances in the ability to address EOCs and the potential for regulatory requirements to make the need to address EOCs more important, it is recommended that future HRA/PRAs identify and model, to the extent necessary, potentially important EOCs. For any EOCs modeled, the guidance given in this report for either pre-initiators or post-initiators are applicable for identifying, modeling, and quantifying EOCs. That is, the same good practices apply whether the error is one of omission or commission.	6.1.3.1
	GP 2: As a Minimum, Search for Conditions that May Make EOCs More Likely. The use of risk in any issue assessment should at least ensure that conditions that promote likely EOCs do not exist. For example, it should be ensured that such conditions have not been introduced by a plant change or modification, or that the plant is not more susceptible to EOCs under the unique set of conditions being examined. When considering the potential for situations that may make EOCs somewhat likely, the premise of any evaluation should be that (1) operators are performing in a rationale manner (e.g., no sabotage), and (2)the procedural and training guidance is being used by the crew based on the plant status inputs they are receiving. Using this premise, EOCs are considered to be largely the result of problems in the plant information/operating crew interface (wrong, inadequate information is present, or the information can be easily misinterpreted) or in the procedure-training/operating crew interface (procedures/training do not cover the actual plant situation very well because they provide ambiguous guidance, no guidance, or even unsafe guidance for the actual situation that may have evolved in a somewhat unexpected way). In either case, significant mismatches can occur between the scenario conditions and the crew's understanding of those conditions. Such mismatches should be searched for and their potential for leading to EOCs should be examined.	6.1.3.2

Analysis Activity	Good Practice	Section
HRA Documentation	GP 1: Document the HRA. The HRA should be documented well enough to allow a knowledgeable reviewer to understand the analysis to the point that it can be at least approximately reproduced and the resulting conclusion reached, if the same methods, tools, data, key assumptions, and key judgments and justifications are used. Hence, the documentation should include the following, but only to the extent it is applicable for the application. The overall approach and disciplines involved in performing the HRA, including to what extent talk-throughs, walkdowns, field observations, and simulations were used; summary descriptions of the HRA methodologies, processes, and tools used; assumptions and judgments made in the HRA, their bases, and their impact on the results and conclusions; the PSFs considered (for at least each of the HFEs important to the risk decision to be made), the bases for their inclusion, and how they were used to quantify the HEPs, along with how dependencies among the HFEs and joint probabilities were quantified; the sources of data and related bases or justifications for the screening and conservative values, and the best estimate values, along with their uncertainties with related bases; the results of the HRA including a list of the important HFEs and their HEPs; and the conclusions of the HRA.	7.1.3.1

3. HRA TEAM FORMATION AND OVERALL GUIDANCE

If the PRA is to realistically include human actions, the modeling of human interactions must consider each action evaluated in the context of a complete accident scenario or sequence of events. To do so, HRA has evolved from the days when PRA analysts sometimes provided the human events of interest to an HRA specialist, who then assigned HEPs to the human events. Such a process is no longer typical nor considered good practice. Understanding an accident sequence context is a complex, multifaceted process. The interaction of plant hardware response and the response of plant operators must be investigated and modeled accordingly. The following characteristics (among others) need to be understood and reflected, as necessary, in the model of a specific human action or group of actions:

* plant behavior and conditions
* timing of events and the occurrence of human action cues
* parameter indications used by the operators and changes in those parameters as the scenario proceeds
* time available and locations necessary to implement the human actions
* equipment available for use by the operators based on the sequence
* environmental conditions under which the decision to act must be made and the actual response must be performed
* degree of training, guidance, and procedure applicability

Much of the guidance in this report is aimed at good practices for understanding the context associated with each modeled human action, and how that context affects both the definition of HFEs and an assessment of their probabilities. It should be noted that in addition to the above characteristics, organizational influences are part of understanding the full context associated with human performance. While this document addresses some related contextual factors (such as staffing resources and administrative controls and biases), it does not specifically address organizational influences as part of the context of human performance. Although work has been done in this area [e.g., Davoudian, Wu, and Apostolakis (Refs. 20 and 21)], the treatment of these influences — and particularly how they are accounted for in estimating human error probability — are still subject to research. Further work is needed to clarify how to incorporate these potentially important influences and to gain experience and credibility from applications.

This emphasis on the need to adequately understand and address context in order to more realistically address human performance is based on advances in our understanding of the factors that can influence human performance. These advances come from recent reviews of operational events involving serious accidents (e.g., ATHEANA, Ref. 11), from related work in HRA [e.g., Hollnagel (Ref. 14) and Singh, Parry, and Beare (Ref. 22)] and other international efforts and recent research in the cognitive sciences that together have provided a clearer picture of the ways in which various factors and situations can interact to influence the occurrence of inappropriate human actions [e.g., Reason (Ref. 23), Woods (Ref. 24), and Endsley (Ref. 25)]. Improvements have been made in how to address the broad range of potential influences on human performance, on both the identification of the human actions to be modeled in the PRA as well as what to consider during screening and detailed quantification of the actions. The guidance in this report provides good practices that reflect these improvements and ensures the proper treatment of context in performing a reasonably realistic HRA.

Identifying the appropriate participants for the HRA (i.e., forming the HRA team) and encouraging the use of certain techniques throughout the HRA process, to ensure that the context associated with each human action is both properly understood and reflected in the PRA, is an activity discussed below with its associated good practices.

3.1 HRA Team Formation and Techniques for a Realistic Analysis

3.1.1 Objective

The objective of HRA team formation and techniques for a realistic analysis is to provide a strong underlying basis to ensure the context for each human action is adequately determined. Properly determining the correct context associated with each modeled human action should lead to HRA results that are credible.

3.1.2 Regulatory Guide 1.200 Position

There is no specific guidance on team formation. However, because of past practices whereby HRA was sometimes treated as an isolated task or for which much of the analysis was performed without, for instance, examining the locations involving the actions or obtaining operators' and trainers' perspectives on the specific scenarios, the following good practices are warranted to stress their importance. Further, while there are references to the use of talk-throughs and simulations in the regulatory guide and its endorsement of the ASME Standard (Ref. 5), the need to use such techniques is emphasized here to confirm or expand any judgment or assumption associated with understanding the context of a human action (e.g., not only to verify the time it takes to perform an action).

3.1.3 Good Practices

3.1.3.1 Good Practice #1: Perform a Multi-Disciplinary, Integrated Analysis

The HRA should be an integral part of the PRA (not performed as an isolated task in the PRA process) whereby the inputs from the following disciplines are used together to define the PRA structure, including which human events need to be modeled, how they are defined and modeled in the PRA, and the considerations used to quantify the associated HEPs:

- PRA modelers
- HRA practitioners (i.e., someone trained or experienced in HRA)
- human factors specialists
- thermal-hydraulics analysts
- operations, training, and maintenance personnel
- other disciplines (e.g., structural engineers, system engineers) as necessary (e.g., structural engineers if the timing of an action is dependent on when and how the containment might fail)

Each discipline provides a portion of the context knowledge. Only when the context is sufficiently understood can HFEs be realistically modeled and quantified.

3.1.3.2 Good Practice #2: Perform Field Observations and Discussions

Besides the review of plant documents, the HRA is performed using the insights gained from the activities listed below. Performance of the activities allows analysts to confirm judgments and assumptions made from the document review and helps them obtain a more well-informed understanding of the context for the various actions and scenarios. If these activities are not performed, there will be uncertainties associated with the context that could be important. Thus, such activities should only not be performed when a generic, non-detailed analysis is deemed adequate for the issue being examined (e.g., the NRC's accident sequence precursor analyses). The activities include the following:

- Walkdowns and field observations of areas where decisions and actions are to take place to understand the equipment involved including the need for any special tools; the plant layout including review of such issues as equipment accessibility, use (or not) of mimic boards, instrumentation availability, labeling conditions, etc.; whether any special fitness needs are required; the time required to reach the necessary locations and perform the desired actions; and the environment in which the actions will need to be performed (e.g., nominal, radiation-sensitive, high-temperature, etc.).

- Talk-throughs of scenarios and actions of interest with plant operators, trainers, or maintenance staff should be performed. Such talk-throughs should include a review of procedures and instructions to learn about the potential strengths and weaknesses in the training and procedures relevant to the actions of interest, identifying possible workload or time pressure or other high-stress issues, identifying potential complexities that could make the desired actions more difficult, and learning of any training biases that may be important to the actions of interest, etc. In addition, Sträter and Bubb (Ref. 26) pointed out that in identifying and searching for EOCs, it is important to obtain a good understanding of operators' intentions in a given scenario. Clearly, inappropriately developed intentions could lead crews to take inappropriate actions (e.g., terminate safety injection). Talk-throughs with crews and trainers will be an ideal time to obtain an understanding of their expected intentions in given scenarios.

- Simulator exercises as a means to observe "near real" crew activities. While it is realized that simulator exercises may not always be possible, it is good practice that at least a representative set of scenarios for the issue under investigation be simulated and observed by the HRA team. In addition to allowing analysts to obtain scenario-specific and related timing information relevant to implementing certain procedure steps, it also allows them to see how plant crews perform as a team and implement their procedures. This could lead to identification of important crew characteristics such as clarity of communications (e.g., whether direction and feedback is clear or potentially ambiguous), the degree of independence that is allowed among individual crew members (e.g., what actions can be performed without general crew knowledge and the extent to which review occurs to ensure that the appropriate actions were taken), the level of aggressiveness followed by the crew (e.g., whether some actions can be and are typically implemented out-of-sequence of the normal procedure step-by-step flow), etc. Moreover, observation of simulator exercises also provides a basis for discussions with operators and trainers about both the scenarios that are observed and those that are unobserved as a result of time or resource constraints.

3.1.4 Possible Impacts of Not Performing Good Practices

Not meeting the above good practices may have little to no impact on the HRA, or it may have a significant impact. Generally, for human actions for which a very simple and straightforward context can reasonably be assumed, no impact is likely. For those actions for which a more complex context is likely (e.g., scenarios with multiple failures [equipment and/or human actions] or relatively unexpected scenarios), there could be a much greater impact (e.g., grossly underestimating the failure probability for a human action). While there is no direct way to "measure" the effect of not performing the above good practices, the implication is that the context used to model and quantify a human action *may not* be correct or complete enough if the above team participants and confirmation techniques are not used at appropriate stages during the analysis. In other words, the modeling and quantification of the human action may not accurately reflect "the current design and operating practices" as discussed in RG 1.200.

The impact, if any, will be manifested via the other good practices such as not addressing all the appropriate performance-shaping or recovery factors, or in results that do not make sense. If this occurs, the source of the insufficiency may be that the good practices in this section were not met (i.e., the team makeup and participation at various stages of the analysis were inadequate or the information gathering and confirmation techniques were not used, leading to incorrect judgments or assumptions about the context). For this reason, these good practices have been provided as an adjunct to all the other good practices in this report, since they provide an underlying basis for the success of the entire HRA process.

4. PRE-INITIATOR HRA

The NRC staff has stated its positions on the ASME Standard (Ref. 5) in Appendix A to RG 1.200 (Ref. 4). The standard separates its requirements into two broad classifications, including (1) those that address the modeling of failures of pre-initiator human actions, and (2) those that address the modeling of failures of post-initiator human actions. This section provides good practices for implementing the requirements for addressing pre-initiator HFEs in a PRA.

Pre-initiator HFEs are events that represent the impact of human failures committed prior to the initiation of an accident sequence (e.g., during test or maintenance or the use of calibration procedures). They are important to model because plant personnel can make the equipment needed to mitigate a particular accident sequence unavailable, thereby reducing the overall capability to respond to the initiating event. Hence, depending on the issue being addressed, this impact may need to be included in a PRA if a detailed assessment of risk is required.

The good practices are categorized under the following four major analysis activities for pre-initiator HRA:

(1)   Identify activities that have the potential to result in pre-initiator human failures

(2)   Screen out activities for which HFEs do not need to be modeled.

(3)   Model specific HFEs corresponding to the unscreened activities.

(4)   Quantify the corresponding HEPs for the specific HFEs.

4.1   Identifying Potential Pre-Initiator HFEs

4.1.1   Objective

The objective is to identify from routine plant actions, those pre-initiator human actions of which failure to perform correctly could result in the human-induced unavailability of PRA-modeled equipment that is credited (i.e., modeled, including the likelihood of either success or failure) in the PRA accident sequences. This is important because these actions represent other potential modes of unavailability of the credited equipment (besides the equipment simply failing to start, or other failure modes in the PRA) that contribute to overall plant risk. Note that not all of the identified actions will be modeled since some may be screened from further analysis in the following analysis activity (screening). The following provides good practices for identifying potential pre-initiator human failures while implementing RG 1.200 (Ref. 4) and the related requirements of the ASME Standard (Ref. 5).

4.1.2   Regulatory Guide 1.200 Position

The ASME Standard calls for a systematic process to be used to identify routine activities that if not correctly completed, may impact the availability of equipment addressed in the PRA. There are multiple supporting requirements in the Standard under high-level requirement HLR-HR-A that address the need to consider test and maintenance activities, calibration activities, and actions that could affect multiple equipment. The regulatory guide states that the NRC staff has no objections to the ASME Standard requirements covering this activity.

4.1.3 Good Practices

4.1.3.1   Good Practice #1: Review Pre-Initiator Procedures, Actions, and Equipment

The main goal of this good practice is to identify pre-initiator human actions with the potential to leave important equipment unavailable. Since the analysis of the actions will be an iterative process throughout the PRA, at this stage the HRA process should include reviews of the items listed below to the extent that is necessary to determine which actions should be included through the screening phase and to obtain the information necessary to apply the screening rules if screening is performed (Section 4.2). The sources of information can be returned to later, if needed to support further analysis. The review should cover the following:

- All routine (scheduled) testing and maintenance, as well as calibration procedures that affect equipment to be credited in the PRA [for core damage frequency (CDF) and LERF], should be identified and reviewed.

- Unscheduled maintenance acts that lead to reconfiguration f a system are opportunities for restoration errors and should be addressed in the analysis.

- Actions specified in the above procedures that realign equipment outside their normal operation or standby status, or otherwise could detrimentally affect the functionality of credited equipment if not performed correctly (e.g., miscalibration) should be identified.

- "Affected" equipment that is routinely acted on and credited in the PRA, which should include the following:
  - the primary systems, structures, and components (SSCs) (e.g., emergency core cooling systems' components, containment cooling systems' components)
  - support systems (e.g., power, air, cooling water)
  - cascading effects among the equipment (e.g., if the realignment of an equipment item in one procedure such as an air-operated valve would implicitly require the subsequent realignment of another equipment item such as isolation of an air line that would then disable a portion of the air system)
  - instrumentation (e.g., indicators, alarms, sensors, logic devices) and controls (e.g., hand switches) that (1) affect automatic operation of the above primary and support system equipment, or (2) is the sole instrumentation relied upon (as opposed to multiple, redundant items) to credit post-initiator human actions to be included in the model (e.g., a single subcooling indication relied upon to meet an emergency core cooling termination criteria which if miscalibrated could induce failure of the appropriate post-initiator operator action).

In addition, a review of operational plant experience should be performed to ensure that potentially important pre-initiators were not missed. The use of relevant generic databases, as well as plant-specific experience, may be appropriate for this review; for example, international stakeholders have indicated the successful use of the International Common Cause Data Exchange (ICDE) database (Refs. 27–29) for helping to identify and understand pre-initiators (e.g., human failure mechanisms and root causes).

4.1.3.2   Good Practice #2: Do Not Ignore Pre-Initiators

The identification process should identify pre-initiator human actions even if they may be potentially covered by the affected equipment failure data. (See section 4.1.4 for additional information.)

4.1.3.3   Good Practice #3: Examine Other Operational Modes and Routine Actions

If applicable and credited in the analysis, the identification process should address other operational modes and routine actions affecting barriers and other structures such as fire doors, block walls, drains, seismic restraints, etc.

4.1.3.4   Good Practice #4: Identify Actions Affecting Redundant and Multiple Diverse Equipment

The identification process needs to include possible pre-initiator actions *at least within each system* where redundant or multiple diverse equipment can be affected by (1) a single action (e.g., misalignment of a valve affecting multiple system trains or even multiple systems), or (2) through a common failure with similar multiple actions (e.g., miscalibrating multiple sensors as a result of incorrect implementation of the same calibration procedure or use of the same miscalibrated standard). For the latter case, the analyst should not duplicate that already covered under the common cause failure modeling of the equipment, but should include consideration of at least the following possible commonalities in deciding whether to include a particular action:

- same crew, same shift performing the actions (common "who" mechanism)

- common incorrect calibration source (common "what" mechanism)

- common incorrect tool, process, or procedure/training, or inadequate material (e.g., wrong grease) (common "what/how" mechanisms)

- proximity in time and/or space/location of similar multiple actions (common "when/where" mechanisms)

- common cues (e.g., same indications, labels, alarms, procedures, steps) (common "why" mechanism)

The more these commonalities exist concurrently with respect to a particular action being considered for initial inclusion in the model, the more the identification process should consider the action as a potentially important pre-initiator action to be included.

4.1.4 Possible Impacts of Not Performing Good Practices and Additional Remarks

Failure to perform the above good practices could lead to an incomplete list of routine human activities that if not completed correctly, may impact the availability of equipment addressed in the PRA. Therefore, this could result in potentially missing (i.e., not modeling) a risk-significant pre-initiator human action. The following related observations should be considered:

- Missing or unnecessarily including a human action is often not a serious mistake (i.e., would not significantly affect the overall risk) unless the human action (1) can affect multiple equipment items, or (2) can affect a single equipment item with a high operating reliability. This is because with common nuclear plant practices and designs, typically those human actions that could affect multiple trains of equipment tend to be the more significant pre-initiator human failures. Those affecting only one equipment item are usually not important unless the equipment item has a high operating reliability (e.g., failure to start or run is in the probability range of 0.0001 or lower) and so the pre-initiator failure probability from failing the human action could be a significant contributor to the unavailability of the equipment.

- The possible human action failures associated with routine test and maintenance or calibration procedures should be included when they could affect critical instrumentation, diagnostic devices, or specific items like pushbuttons, etc. that have no redundancy or diverse means of function. While typically such situations do not exist in NPPs, changes to the plant could conceivably and unintentionally create such a situation. Affecting the operator's ability to take the desired action is similar, functionally, to affecting the equipment item itself which is to be activated. Hence, the analysis approach should ensure that such situations, from a possible pre-initiator perspective, do not exist or if they do, they are addressed.

- In practice it is best to include pre-initiator human actions even if the associated failure already may be included in the failure data for the affected equipment item (e.g., in the failure-to-start data). This is because it is often hard to determine if the failure databases include such human failures since data bases are typically insufficiently documented to know if the potential pre-initiator failure is already included. Generally, unless the failure can affect multiple equipment items, either missing the failure or double-counting the failure have small effects on the outcome of the PRA. Potential double-counting is the most conservative thing to do, and yet typically not a serious overestimation of the failure's significance. In addition, including all identified pre-initiators gives analysts the opportunity to identify the significance of potentially problematic actions such as those with procedural or training problems, those that do not require appropriate checks, etc.

- If applicable, the possible failures associated with routine test and maintenance or calibration procedures should be included when they could affect equipment critical to external events such as fire barriers (e.g., opening a fire door and failing to restore it to a closed position), seismic restraints, floor drains and barriers, wind barriers, etc. While typically such situations do not exist in NPPs since such equipment items often do not have routine test, maintenance, or calibration activities that would adversely affect their function, changes to the plant or plant practices, for instance, could conceivably and unintentionally create such a situation. To the extent the analysis assumes the functionality of these normally highly reliable devices, pre-initiator failures that could affect these devices could be potentially important. Hence, the analysis approach should ensure that such situations, from a possible pre-initiator perspective, do not exist or if they do, they are addressed.

- Considering the potential importance of actions that affect multiple equipment items, the identification process should search for actions that affect multiple equipment items *at least within a system* (e.g., auxiliary feedwater system, reactor core injection system) as this represents the current state-of-the-art in PRA. A search across multiple systems (e.g., auxiliary feedwater and high-pressure injection) is an expansion of the current state-of-the-art and should not be expected except for those cases where the same instrumentation or equipment (e.g., pressure signals, same tank level equipment) activates or affects multiple systems.

4.2 Screening Activities for Which HFEs Do Not Need To Be Modeled

4.2.1 Objective

The objective of screening is to eliminate from consideration (if desired) those activities for which associated failures (HFEs) do not need to be modeled because they should be probabilistically unimportant. The screening process, although largely qualitative, is based on the belief that certain design or operational practices make some pre initiator failures sufficiently unlikely that they will not be risk-significant failures and, therefore, do not need to be modeled (although there may be other reasons to maintain these activities in the model as discussed in Section 4.2.4). The following subsection provides good practices for screening out pre-initiator human actions and associated human failures while implementing RG 1.200 (Ref. 4) and the related requirements of the ASME Standard (Ref. 5).

4.2.2 Regulatory Guide 1.200 Position

The ASME Standard addresses allowable screening of activities based on practices that limit the likelihood of errors in those activities. There are multiple supporting requirements in the ASME Standard under high-level requirement HLR-HR-B that address screening rules or criteria, as well as the requirement to not screen actions that could affect multiple equipment items. The regulatory guide states that the NRC staff has no objections to the ASME Standard requirements covering this activity.

4.2.3 Good Practices

4.2.3.1   Good Practice #1: Screen Pre-Initiators with Acceptable Restoration Mechanisms or Aids

A candidate pre-initiator action can be screened out (i.e., not to be modeled) if the nature of the associated action meets any one of the following criteria, and the reason for screening is documented (see exception under Good Practice #2 below):

- The affected equipment will receive an automatic realignment signal and it can respond if demanded (i.e., the equipment will not have been disabled by the human actions).

- There is a valid post-maintenance test/functional check (a test or functional check that has been shown to work consistently) after the original manipulation which will reveal misalignment or incorrect status (e.g., faulty position, improper calibration).

- Following the original action(s), there is an independent second verification of equipment status that uses a written checklist that will verify incorrect status.

- There is a valid check (one that has been shown to work consistently), at least once per shift, of equipment status that will reveal misalignment or incorrect status.

- There is a compelling signal (e.g., annunciator or indication) of improper equipment status or inoperability in the control room, it is checked at least once per shift or once per day, and realignment can be easily accomplished.

- Other criteria apply, as long as it can be demonstrated, using an acceptable model such as the Technique for Human Error Rate Prediction (THERP, Ref. 11) or the Accident Sequence Evaluation Program (ASEP, Ref. 30) that the resulting HEPs would be low compared with the failure probabilities (e.g., failure to open) of the equipment.

4.2.3.2   *Good Practice #2: Do Not Screen Actions Affecting Redundant or Multiple Diverse Equipment*

Do not screen those actions and possible pre-initiator failures that simultaneously affect multiple (redundant or diverse) equipment items (see Good Practice #4 under Section 4.1.3).

4.2.3.3   *Good Practice #3: Reevaluate the Screening Process for Special Applications*

For a specific PRA application and depending on the issue being addressed (e.g., examination of a specific procedure change), revisit the original PRA screening process to ensure issue-relevant human actions have not been deleted from the PRA prior to its use to assess the new issue.

4.2.4 Possible Impacts of Not Performing Good Practices and Additional Remarks

Failure to perform the above good practices could lead to inappropriate screening out of human actions and therefore not including the actions in the PRA. This could result in potentially missing a risk-significant pre-initiator human action. The following related observations should be considered.

- Generally, screening out pre-initiator human failures (i.e., don't have to be modeled) is acceptable based on experience with past PRAs and the types of pre-initiator failures that are typically found to be unimportant. This is done to simplify the model and not expend resources addressing unimportant pre-initiator human actions. It should be clear that an appropriate level of investigation has been performed to ensure the above criteria have been met and if these or other criteria are used, their justification is documented for outside review. It is advisable to keep a record of all screened actions for later reference when performing specific applications (see Good Practice #3). When in doubt, it is recommended the pre-initiator action not be screened out, but the corresponding human failure modeled in the PRA for further analysis.

- Since pre-initiator human actions and related failures affecting multiple equipment items can sometimes be risk-important, none of these should be screened out but should be modeled and examined in more detail in the PRA because of the potential consequences of the failure.

- There can be a tendency to use an existing PRA model to address issues, such as changes to the plant, without spending the appropriate time to revisit some of the underlying assumptions and modeling choices made to create the original PRA. However, such a review should be done to see if these assumptions and choices still apply for the issue being addressed. Some pre-initiator human failures may not have been included in the original PRA (i.e., screened out) that in light of the new issue being addressed, should now be included in the model (i.e., could be important for addressing the issue). Hence, it is good practice to implement a process that determines whether some of the formerly screened out pre-initiator human failures should be added back into the model in order to appropriately address the issue.

- In spite of the above, it may be desirable to not eliminate any of the pre-initiator actions and the related possible HFEs from the PRA model. This will allow the greatest flexibility for future applications, including avoiding the need to revisit whether "screened out" actions need to be put into the model to investigate a particular issue. For instance, extending the PRA model to low-power/shutdown applications where an automatic realignment may be defeated, can be more easily analyzed if the potential error, even with automatic realignment, is already in the model (rather than having to put it into the model to investigate the shutdown issue). Since most of the work to identify and investigate the nature of the activities will have already been performed, including failures of these activities in the model should require minimal additional effort. The use of conservative or otherwise less-detailed evaluations can be used to minimize this effort. It should be clear that RG 1.200 and the ASME Standard allow screening such pre-initiator events and provide guidance on the screening criteria, which are elaborated upon here. Screening is an acceptable practice. However, for the reasons cited here, analysts should evaluate the usefulness of this screening activity considering the future and broader uses of the PRA model. Further, not screening such events avoids the need to document and justify the acceptability of the screening process.

4.3   Modeling Specific HFEs Corresponding to Unscreened Human Actions

4.3.1   Objective

The objective is to define how the specific pre-initiator HFE is to be modeled in the PRA to accurately represent the failure of each action identified and not screened out from the above analysis activities. The HFE needs to be linked to the affected equipment (single or multiple) and needs to appropriately define the failure mode of that equipment that makes the equipment unavailable. The following provides good practices for modeling pre-initiator HFEs while implementing RG 1.200 (Ref. 4) and the related requirements of the ASME Standard (Ref. 5).

4.3.2 Regulatory Guide 1.200 Position

The ASME Standard calls for the modeling of pre-initiator HFEs based on the impact of the mode of unavailability of the affected equipment item(s) in the PRA. There are multiple supporting requirements in the Standard under high-level requirement HLR-HR-C that address the modeling level of detail for each HFE and the modes of failure (as unavailabilities) to be considered. The regulatory guide states the NRC staff has no objections to the ASME Standard requirements covering this activity.

4.3.3 Good Practices

4.3.3.1 *Good Practice #1: Include HFEs for the Unscreened Human Actions in the PRA Model*

Define each specific pre-initiator HFE to be modeled in the PRA as a basic event that describes the human-induced failure mode and locate the HFE in the model such that it is linked to the unavailability of the affected component, train, system, or overall function (i.e., level of modeling) depending on the effect(s) of the HFE (e.g., a single valve will not close, a train will be inappropriately isolated, the automatic start signal for an entire system will be disabled). The following considerations, as a minimum, should be taken into account when defining the pre-initiator failure level properly in the PRA:

- whether the nature of the manipulation affects an entire train, system, etc., making it logical to define the HFE at the level of an entire train, system, etc.

- whether multiple individual actions affecting multiple pieces of equipment (e.g., different components) can be combined as a single pre-initiator HFE affecting a higher level of equipment resolution (e.g., the train containing the different components). This can be done as long as the following criteria are met:
  - ▸ The actions and effects are related.
  - ▸ The factors affecting the quantification of the single HFE will not be significantly different than those that would have been relevant for the individual actions (e.g., the same PSFs will be relevant, as is discussed later) or the quantification result will be conservatively bounding compared to modeling and quantifying the individual actions separately.
  - ▸ There are no potential commonalities/dependencies with other pre-initiator actions elsewhere in the model so that potential common failures among similar individual actions might be missed (e.g., miscalibration of multiple signal channels).
  - ▸ The level of detail already modeled in the PRA (e.g., train, system) for failures of the associated equipment (but note that this factor is less important than those above).

The failure modes (fail to close, fail to start, etc.) or modes of unavailability (valve left in wrong position, etc.) should be a direct result of considering the equipment affected and the effects of the human-induced failure (refer to all the Good Practices under Section 4.1.3), and stem from failure to restore equipment and/or otherwise correct the adverse effect (such as miscalibration) so that the equipment is again operable. The failure modes should clearly describe the HFE effect to ensure proper interpretation of the HFE in the model (e.g., only two of three redundant sensors need to be disabled to make the actuation signal unavailable, and not all three sensors have to be disabled).

As an aid to ensure appropriate modeling, it is a recommended practice (but not necessary) that the pre-initiator failure be placed in proximity, in the PRA model, to the equipment affected by the human failure. In this way, a quick comparison can be made between the equipment failure and the pre-initiator human failure to ensure they are consistent.

4.3.4 Possible Impacts of Not Performing Good Practices and Additional Remarks

Failure to either (1) properly link the pre-initiator human failure to the affected equipment item(s), or (2) properly model the effect by the appropriate failure mode of the equipment item(s) will likely lead to misrepresentation of the unavailability of the equipment caused by a pre-initiator human failure in the PRA. Depending on the specific misrepresentation, the risk effect of the human failure could be overemphasized (such as if the human failure is modeled as affecting more equipment items than it actually does or it is modeled as causing a failure mode that it does not cause), underestimated (such as if the human failure is modeled as affecting fewer equipment items than it actually does or a failure mode that is caused by the human failure is missed in the model), or even missed entirely. This can result in inaccuracies in the PRA results and particularly incorrect assessments of the importance of pre-initiator human failures. The precise definition of the pre-initiator basic events and their placement in the model (from both a logic and failure mode standpoint) ultimately defines how the model addresses the effects of the human failures. This needs to be done accurately if the model is going to logically represent the real effects of each human failure and if the corresponding HFE is going to be correctly quantified (as discussed later).

4.4 Quantifying the Corresponding HEPs for Pre-Initiator HFEs

4.4.1 Objective

The objective is to address how the HEPs for the modeled HFEs from the previous analysis activity are to be quantified. This section provides good practices guidance on an attribute or criteria level and does not endorse a specific tool or technique, although THERP (Ref. 9) or its ASEP (Ref. 30) simplifications are among those often used. Ultimately, these probabilities (along with the other equipment failure and post-initiator HEPs) and initiating event frequencies are all combined to determine such risk metrics as CDF, LERF, ΔCDF, ΔLERF, etc., as addressed in RG 1.174 (Ref. 3). The following provides good practices for quantifying pre-initiator HFEs while implementing RG 1.200 (Ref. 4) and the related requirements of the ASME Standard (Ref. 5).

4.4.2 Regulatory Guide 1.200 Position

The ASME Standard calls for a systematic process for assessing the pre-initiator HEPs that addresses plant- and activity-specific influences. There are multiple supporting requirements in the Standard under high-level requirement HLR-HR-D that address many factors associated with quantifying the HEPs. These include when screening vs. detailed estimates are appropriate, PSFs considered in the evaluations, treatment of recovery, consideration of dependencies among HFEs, uncertainty, and reasonableness of the HRA results. The regulatory guide states the NRC staff has no objections to the ASME Standard requirements covering this activity.

4.4.3 Good Practices

4.4.3.1   Good Practice #1: Use Screening Values During the Initial Quantification of the HFEs

The use of screening-level HEP estimates is usually desirable during the PRA development and quantification, with the estimates preferably assigned once much of the modeling is complete. This is acceptable (and almost necessary since not all the potential dependencies among human events can be anticipated) provided (1) it is clear that the individual values used are overestimations of the probabilities that would be developed if detailed assessments were to be performed **and** (2) dependencies among multiple HFEs appearing in an accident sequence are conservatively accounted for. These screening values should be set so as to make the PRA quantification process more efficient (by not having to perform detailed analysis on every HFE), but not so low that subsequent detailed analysis would actually result in higher HEPs. The screening estimates should consider both the individual events and the potential for dependencies across multiple HFEs in a given accident sequence (scenario). To meet these conditions, it is recommended that (unless a more detailed assessment is performed of the individual or combination events to justify lower values):

*   No individual pre-initiator HEP screening value should be lower than 0.01 (this is typical of the highest pre-initiator values in PRAs, recognizing that the nature of these tasks usually involves the use of familiar procedures, performed under non-stress conditions on a frequent basis).

*   Multiple HFEs in the same sequence should not have a joint probability value lower than 0.005 (accounts for a 0.5 high dependency factor) at this stage.

4.4.3.2   Good Practice #2: Perform Detailed Assessments of Significant HFEs

As needed for the issue being addressed to produce a more realistic assessment of risk, detailed assessments (not only screening estimates) should be performed for at least the significant HFE contributors. [See Table A-1 in RG 1.200 (Ref. 4) for a definition of "significant contributor."] The PRA analyst can define the significant contributors by using typical PRA criteria [not addressed here, but see Section 2.3 of NUREG-1764 (Ref. 8)], such as importance measure thresholds as well as other qualitative and quantitative considerations. While using screening-level values (supposedly purposely conservative) may, at first, seem to be a "safe" analysis process, it can have negative impacts. Screening values can focus the risk on inappropriate human actions or related accident sequences and equipment failures because of the intentionally high HEPs. Such incorrect conclusions need to be avoided by ensuring that the model includes a sufficient set of more realistic, detailed HEPs.

4.4.3.3   Good Practice #3: Revisit the Use of Screening Values vs. Detailed Assessments for Special Applications

For a specific PRA application and depending on the issue being addressed (e.g., examination of a specific procedure change), revisit the use of screening vs. detail-assessed HEPs to ensure issue-relevant human actions have not been prematurely deleted from the PRA or there is an inappropriate use of screening vs. detailed values to properly assess the issue and the corresponding risk.

4.4.3.4　　Good Practice #4:　Account for Plant- and Activity-Specific PSFs in the Detailed Assessments

HEP assessments should account for the most relevant plant- and activity-specific PSFs in the analysis of each pre-initiator HFE.　There is not one consensus list of appropriate contextual factors (e.g., plant conditions, PSFs, activity characteristics, etc.) to be considered in the evaluation of the pre-initiator HEPs.　Additionally, for a specific action, which factors are most relevant may be different (e.g., perhaps one action is time-sensitive because it is done in a high-radiation area while another is most affected by the complexity of steps with many opportunities to make undetected mistakes).　It should be qualitatively apparent that the factors seemingly most relevant to the action (based on an understanding of the action [as might be derived from a sound task analysis]) have been considered in the corresponding HEP estimate.

The following factors are typically important to address because they tend to be variable and not always optimal based on typical nuclear plant practices:

- whether the activity relies on the use of written work plans, job briefs, and related procedures (positive influences tending to lower the HEP) rather than verbal guidance and/or memory (more negative influences tending to raise the HEP), as well as the quality of the information (e.g., look for ambiguities, incompleteness, inconsistencies, etc. that are negative influences and thus tend to raise the HEP)

- whether the activity is complex (e.g., involves multiple and/or repetitive steps that are hard to track, requires coordination of multiple personnel, does not use appropriately human-factored checklists, involves several variables, requires that calculations be performed, requires sensitive adjustments)

- what ergonomic issues (e.g., layout, available information [instruments, alarms, computer readouts, etc.], labeling, readability, physical demands) are relevant

Note that information about such factors can be greatly enhanced by performing talk-throughs and walkdowns of the actions and whenever possible, observing actual crews perform the actions (or at least sample actions).

The following factors tend to be less important either because of typical nuclear plant practices or because the factors are typically less relevant:

- skill level/experience/training of the crew (which is typically adequate in nuclear plants for the jobs each crew member is to perform)

- stress level (which is not usually relevant in pre-initiator failures unless special situations such as potential personal harm, the need for fast sequential responses, etc. play a role)

- environmental factors such as temperature, humidity, radiation, noise, lighting, etc., which are typically sufficiently benign (with the exception of special circumstances such as a high-radiation environment that can lead to the desire to hurry the actions)

- availability of time (which is not usually a strong factor in pre-initiator failures)

Nonetheless, the evaluation should ensure that the typical practice or "irrelevancy" is not compromised.

If the large majority of these factors negatively affect human performance, or even if only one or two have a strong negative influence, the HEP will tend to be higher (e.g., 0.01 to 0.1 or even higher, not accounting for recovery addressed under Good Practice #5 below). Conversely, mostly positive influences should yield lower HEPs (e.g., 0.001, with additional recovery factors still to be applied as addressed under Good Practice #5 below).

4.4.3.5　　　*Good Practice #5: Apply Plant-Specific Recovery Factors*

Each HFE modeled in the PRA should be investigated for opportunities to recover the initial failure, so that the HEP for the HFE can be justifiably reduced when there is an accounting for the recovery potential. This is particularly relevant for those HFEs analyzed using detailed assessments, although it may also be applicable for those HFEs to be assigned screening values. Multiple recoveries may be acceptable, where appropriate. However, any dependencies among the initial failure and possible recovery actions, and among the recovery actions themselves, must be considered (see Good Practice #6 below). In accordance with the ASME Standard (Ref. 5) and as elaborated upon here, typical pre-initiator recovery actions[1] or considerations include the following:

- post-maintenance or post-calibration tests that are required and performed by procedure

- independent verification that uses a written checklist which verifies component status following maintenance/testing/calibration and its practice has been verified by walkthroughs and examination of plant experience

- a separate check of component status that is made by the original performer at a later time and that uses a written checklist

- work shift or daily checks that are performed for component status and uses a written checklist

- compelling feedback (e.g., alarm) that supports detection and quick recovery of the original failure

- combinations of the above

The more of these considerations that are applicable for a given pre-initiator HFE, the more the situation tends to increase the recovery potential (i.e., decrease the HEP). To the extent they are independent, each recovery can result in a multiplier (e.g., 0.1) on the original HEP estimate thereby reducing its overall value. However, the analyst should know that there is a logical limit to such reductions. For example, credit for two checks should not be given if there are dependencies (e.g., same or similar job aids and cues) between the checks done by the original performer and the independent verifier.

---

[1]　Note that the definition of a recovery action and its distinction from a repair action has been adopted from RG 1.200 (Ref. 4). *Recovery action* is defined as: a PRA modeling term representing restoration of the function caused by a failed SSC, by bypassing the failure. Such a recovery can be modeled using HRA techniques regardless of the cause of the failure. *Repair* is defined as a general term describing restoration of a failed SSC by correcting the failure and returning the failed SSC to operability. HRA techniques cannot be used since the method of repair is not known without knowing the specific causes.

Basic HEPs for pre-initiator HFEs for nuclear plant applications (including recovery) are typically expected in the range of 0.01 (among the highest) to 0.0001. Any values below the range of 0.0001 to 0.00001 should be considered suspect unless justified.

4.4.3.6 Good Practice #6: Account for Dependencies Among the HEPs in an Accident Sequence

Dependencies among the pre-initiator HFEs, and hence the corresponding HEPs in an accident sequence, should be quantitatively accounted for in the PRA model. This is particularly important so that combined probabilities are not inadvertently too optimistic, resulting in the inappropriate decrease in the risk-significance of human actions and related accident sequences and equipment failures. In the extreme, this could result in the inappropriate screening out of accident sequences from the model because the combined probability of occurrence of the events making up an accident sequence drops below a threshold value used in the PRA to drop sequences from the final risk results.

To address these dependencies, usually a level or degree of dependence among the HFEs in an accident sequence is determined, at first qualitatively (e.g., low, high, complete), and then combined HEPs are assessed accordingly. Once the first HEP has been estimated, subsequent quantitative factors for dependent human failures or recoveries of the original failure are typically expected to be:

- 0.01 to 0.1 for low dependence
- 0.1 to 0.5 for high dependence
- >0.5 for very high dependence
- 1.0 for complete dependence

Note that specific tools/techniques may use somewhat different probabilities than provided herein based on specific considerations. The values above provide at least a rough level of expectation for staff who are not extremely familiar with the specific tools and the values that are typically assigned.

In establishing the level of dependence, Good Practice #4 under Section 4.1.3 addresses typical commonalities that tend to make HEPs more dependent. If, for example, an HFE is not independent of another HFE, then once the first human failure occurs, there is an increased likelihood that a similar second or third, and so on, human failure will also occur. For example, failing to restore the lineup of one train of equipment after a test may increase the likelihood of failing to restore the second train of equipment after a similar test. Good Practice #5 (above) addresses recovery characteristics that tend to break up these commonalities because they "recover" any initial error, making the individual HFEs more independent. The more the types of commonalities addressed under Good Practice #4 under Section 4.1.3 exist and the less corresponding recoveries under Good Practice #5 above exist, the higher should be the assessed level of dependence among the HFEs. To the extent the converse is true, low or even no dependence should be assessed.

4.4.3.7 Good Practice #7: Assess the Uncertainty in HEP Values

Point estimates should be mean values for each HEP (excluding screening HEPs), and an assessment of the uncertainty in the HEPs should be performed (at least for the significant HEPs) to the extent that these uncertainties need to be understood and addressed in order to make appropriate risk-related decisions. Assessments of uncertainty are typically performed in any or all of the following ways:

- developing uncertainty distributions for the HEPs

- propagating them through the quantitative analysis of the entire PRA, such as by a Monte Carlo technique

- performing sensitivity analyses that demonstrate the effects on the risk results for extreme estimates in the HEPs based on at least the expected uncertainty range

Note that, in some cases, it may be sufficient to address the uncertainties only with qualitative arguments without the need to specifically quantify them (e.g., justifying why the HEP cannot be very uncertain or why a change in the HEP has little relevance to the risk-related decision to be made).

When developing specific uncertainty distributions, the uncertainties should include (1) those epistemic uncertainties existing because of lack of knowledge of the true expected performance of the human for a given context and associated set of PSFs, and (2) consideration of the combined effect of the relevant aleatory (i.e., random) factors *to the extent they are not specifically modeled in the PRA* and to the extent that they could alter the context and PSFs for the HFE. For pre-initiator HFEs, there should be few or no aleatory factors worthy of consideration, since typically the procedure used, the environment experienced, etc., do not randomly change (at least significantly). But, for example, if different and significant crew experience levels are known to exist, it is random as to which crew will perform the pre-initiator action at any given time. In such a case, the analyst could choose to explicitly model the HFE as two different events, each with a unique HEP; one for when a less experienced crew is on shift and one for when a more experienced crew is on shift. If this is not done, the mean for the single HFE/HEP should represent the average crew experience level, and the uncertainty should reflect the uncertainty attributable to crew experience and any other relevant factors. Again, aleatory factors are typically not very relevant to pre-initiator HEPs and so typically are not important to address.

Whatever uncertainty distributions are used, the shape of the distribution (e.g., log-normal, beta, etc.) is typically unimportant to the overall risk results (i.e., the results are usually not sensitive to specific distributions). Further, typical uncertainties include values for the HEP that represent a factor of 10 to 100 between the lower bound value and the upper bound value that encompass the mean value. However, it should be noted that some distributions, e.g., log-normal, can give probabilities greater than 1.0 for HEPs that are relatively high.

4.4.3.8   Good Practice #8: Evaluate the Reasonableness of the HEPs Obtained Using Detailed Assessments

The pre-initiator HEPs (excluding the screening HEPs) should be reasonable from two standpoints:

*   first and foremost, relative to each other (i.e., the probabilistic ranking of the failures when compared one to another)

*   in absolute terms (i.e., each HEP value), given the relative strengths of the positive and negative PSFs identified as being important and the presence or absence of recovery factors

Such reasonableness should be checked based on consideration of actual plant experience and history, against other evaluations (such as for similar actions including similar PSFs and context at other plants), and the qualitative understanding of the actions and the relevant contexts and PSFs under which the actions are performed.

It is suggested that a rank-ordered list of the pre-initiator HFEs by probability be used as an aid for checking reasonableness. For example, simple, procedure-guided, independently checked actions should have lower HEPs than complex, memorized, not checked actions, all other factors being the same. Similarly, with respect to the reasonableness of the absolute values, HFEs with many positive PSFs or several strong recovery factors could have values in the range of 0.0001 to 0.00001, while HFEs with dominating negative PSFs and no independent checks could be in the range of 0.1 to 0.01. Typical expectations of pre-initiator HEPs can be widespread (~0.01 to approximately 0.00001) and depend particularly on the relevant contextual factors, applicable recoveries, and proper consideration of dependencies as discussed under many of the Good Practices covered above.

4.4.4 Possible Impacts of Not Performing Good Practices and Additional Remarks

Failure to quantify the pre-initiator HEPs as realistically as necessary using the good practices articulated above (except for where higher screening estimates are used, purposely and appropriately), could result in improper HEPs and thus inaccuracies in the PRA results and particularly incorrect assessments of the importance of pre-initiator human failures. The risk effect of the human failure could be overemphasized [such as if the human failure is estimated with too high (pessimistic) an HEP or a high screening estimate is used where a more realistic detailed estimate is appropriate], or underestimated [such as if the human failure is estimated with too low (optimistic) an HEP or a dependency among failures is not accounted for resulting in too low a joint probability for multiple human failures]. Besides these concerns about inaccuracies in the HEP quantification and, thus, whether the HEPs "make sense," as well as the resulting potential misinformation about the significant risk contributors if quantification is not done well, the following related observations are noted:

*   Screening is a useful and, most often, necessary part of HRA so as to avoid the expenditure of resources on unimportant human events and accident sequences. The above guidance is aimed at allowing a level of useful screening without inadvertently and inappropriately allowing the analytical phenomenon of, for instance, multiplying three human events in the same sequence each at a screening value of 0.01 to yield a 0.000001 combined probability, without checking for dependencies among the human

events. In such a case some HFEs and combinations of events, or even whole accident sequences, may inappropriately screen out of the PRA model entirely because the accident sequence frequency drops below a model threshold. Hence, some of the significant individual or combination contributors may be missed. This is why the estimated HEP values both individually and for combined events should not be too low during the screening stage. Further, if these estimated screening values are left permanently assigned to some HFEs that should be assessed with more detail to obtain a more realistic assessment of risk (supposedly lowering the probability), the risk-significance of these HFEs and related equipment failures are likely to be overemphasized at the expense of improperly lessening the relative importance of other events and failures.

- It is important to be sure that dependencies among the various modeled HFEs including the associated recoveries, have been investigated (e.g., the same person as the originator of the action performing the recovery may be more prone to fail to detect the original failure than an independent checker). Treating HFEs and any corresponding recoveries as independent actions without checking for dependencies (thereby being able to multiply the individual HEPs) can inappropriately lessen the risk-significance of those HFEs and related equipment failures in accident sequences. This can cause the inappropriate screening out of accident sequences because the sequences quantitatively drop below a model threshold value as discussed above under screening. Proper consideration of the dependencies among the human actions in the model is necessary to reach the best possible evaluation of both the relative and absolute importance of the human events and related accident sequence equipment failures.

- The use of mean values and addressing uncertainties are a part of the guidance in RG 1.174 (Ref. 3) and, to the extent addressed therein, the HRA quantification needs to be consistent with that guidance when making risk-informed decisions. The estimates should reflect, to the extent possible, the as-built and as-operated conditions as addressed in the plant- and activity-specific PSFs.

- There can be a tendency to use an existing PRA model to address issues such as changes to the plant, without spending the appropriate time to revisit some of the underlying assumptions and modeling choices made to create the original PRA. However, such a review should be done to see if these assumptions and choices still apply for the issue being addressed. For instance, some pre-initiator HFEs may be quantified in the original model using a set of screening estimates and detailed failure probabilities that may not be appropriate for the new issue being addressed. As an example, where high estimates as screening HEPs may have been acceptable for purposes of the original PRA, these supposedly conservative values may overestimate the contribution of these HFEs for the issue being addressed. Further, the relative risk contribution of equipment and associated accident sequences with which the HFEs appear, may be artificially too high (and, therefore, other events too low) because of the screening values. Hence, it is good practice to revisit the use of screening and detailed HFE probabilities in order to appropriately address the issue.

5. POST-INITIATOR HRA

The NRC staff has stated their positions on the ASME Standard (Ref. 5) in Appendix A to RG 1.200 (Ref. 4). The standard separates its requirements into two broad classifications, including (1) those that address the modeling of failures of pre-initiator human actions, and (2) those that address the modeling of failures of post-initiator human actions. This section provides good practices for implementing the requirements for addressing post-initiator HFEs in a PRA.

Post-initiator HFEs are events that represent the impact of human failures committed during actions performed in response to the initiation of an accident sequence (e.g., while following post-trip procedures or performing other recovery actions). They are important to model because humans can have a direct influence on the mitigation or exacerbation of undesired plant conditions after the initial plant upset. Hence, depending on the issue being addressed, this impact may need to be included in a PRA if a realistic assessment of risk is required.

The good practices are categorized under the following four major analysis activities for post-initiator HRA:

(1)   Identify potential post-initiator human failures.
(2)   Model specific HFEs corresponding to the human actions.
(3)   Quantify the corresponding HEPs for the specific HFEs.
(4)   Add recovery actions to the PRA.

5.1   Identifying Potential Post-Initiator HFEs

5.1.1   Objective

The objective is to identify the key human response actions that the operators may need to take in response to a variety of possible accident sequences and that will therefore need to be modeled in the PRA. This is important since failures associated with these actions (e.g., failure to start standby liquid control, failure to initiate feed-and-bleed, failure to properly control steam generator feed flow, failure to align containment/suppression pool cooling) are represented in the PRA, such that in combination with equipment failures, are expected to lead to core damage and/or large early releases. Such failures contribute to the overall risk and, thus, a systematic process needs to be followed to identify these response actions. The following provides good practices for identifying post-initiator human failures while implementing RG 1.200 (Ref. 4) and the related requirements in the ASME Standard (Ref. 5).

5.1.2   Regulatory Guide 1.200 Position

The ASME Standard calls for a systematic review to identify operator responses required for each of the accident sequences. There are multiple supporting requirements in the ASME Standard under high-level requirement HLR-HR-E that address what to review as well as the types of actions to be included. Use of talk-throughs and simulator observations are also addressed as part of the supporting requirements (although this is best dealt with under the modeling and quantification analysis activities of the analysis and thus addressed more fully later in this report). The regulatory guide states that the NRC staff has no objections to the ASME Standard requirements covering this activity.

5.1.3 Good Practices

5.1.3.1   Good Practice #1: Review Post-Initiator-Related Procedures and Training Materials

Reviews of the following form the primary bases for identifying the post-initiator actions:

- plant-specific emergency operating procedures (EOPs)
- abnormal operating procedures (AOPs)
- annunciator procedures
- system operating procedures
- severe accident management guidelines (SAMGs)
- other special procedures (e.g., fire emergency procedures), as appropriate
- actual experience responding to operational disruptions, plant trips, etc.

The overall goal of the review is to identify ways operators (crews) are intended to interact with the plant equipment after an initiator. The ways they interact will be a function of the various conditions that can occur, as defined by the development of the PRA accident sequences and associated equipment unavailabilities and failure modes. To meet this goal, analysts should particularly note where operator actions that will directly influence the behavior of the system or affect critical functions are called out in these procedures and under what plant conditions and indications (cues) such actions are carried out (note, some actions may be performed immediately and without regard to the specific situation, while others will be plant status and cue dependent). It will also be useful at this time to examine whether there are any potential accident conditions under which the procedures might not match the situation as well as would be desired (e.g., potentially ambiguous decision points or incorrect guidance provided under some conditions). Information about such potential vulnerabilities will be useful later during quantification and may help identify actions that need to be modeled.

While a walkdown of the control room and observations of simulator exercises and talk-throughs with crews about various accident scenarios are probably most important during the modeling phase, if time and resources allow, they may also be useful during the identification phase to help analysts understand the procedures and how they are implemented by the crews.

To the extent that plant-specific or general industry experience may be useful for identifying potential post-initiator actions of interest, it is recommended that this too be examined to provide a comprehensive set of post-initiator actions.

5.1.3.2   Good Practice #2: Review Functions and Associated Systems and Equipment To Be Modeled in the PRA

The PRA team's plant, system, and operations knowledge is important to the performance of this review and identification task. The process should accomplish the following:

- Identify the functions and associated systems and equipment to be modeled in the PRA.

- Identify whether the function is needed (e.g., injection) or undesired (e.g., stuck-open safety relief valve) for each scenario being addressed, recognizing that the need for a function may vary with different initiators and sequences.

- Identify the systems/equipment that can contribute to performing the function or cause the undesired condition (including structures and barriers such as fire door and floor drains, especially for external event analyses).

- Identify ways the equipment can functionally succeed (i.e., the success criteria) and fail.

Based on the above, the analysts together identify ways the operators are (1) intended/required to interact with the equipment credited to perform the functions modeled for the accident sequences modeled in the PRA, and/or (2) to respond to equipment and failure modes that can cause undesired conditions per the PRA. During the identification process, it is helpful to use action words such as actuate, initiate, isolate, terminate, control, change, etc. so that the desired actions are clear. This process ensures that the logic models correctly model the impact of emergency operating procedures and other procedures, such as AOPs and annunciator response procedures, on the accident sequence development.

5.1.3.3   *Good Practice #3: Look for Certain Expected Types of Actions*

While the specific actions to be identified may be plant-specific, in general, the list below provides the types of actions that are expected to be identified. Note that actions that are "heroic" (e.g., operators must enter an extreme high-radiation environment in order to perform) or that are performed without any procedure guidance or are not trained on, should not be included or credited in the analysis. Exceptions may be justified, but this should not be normal practice. The analysts should include the following considerations:

- necessary and desired/expected actions (e.g., initiate residual heat removal, control vessel level, isolate a faulted steam generator, attempt to reclose a stuck-open relief valve)

- backup actions to failed or otherwise defeated automatic responses (e.g., manually start a diesel generator that should have auto-started), but be sure that the action can be credited to recover the auto-failure mode

- anticipated procedure-guided or skill-of-the-craft recovery actions (e.g., restore offsite power, align firewater backup) although these may best be defined later as the PRA quantification begins and important possible recovery actions become more apparent

- actions for which performance requires permission from other emergency or technical support staff (e.g., some SAMGs and the corresponding interactions with a technical support team)

Consistent with present day state-of-the-art, actions for which failures involve an EOO should be included when identifying post-initiator actions of concern. These involve failures to take the appropriate actions as called out in the procedures and/or trained on or expected as common practice. For example, failure to initiate feed-and-bleed or failure to start standby liquid control, are EOOs. Possible actions for which failure would involve an EOC have generally been beyond PRA practice, but some issues may require that the PRA/HRA address such failures. EOCs involve performing expected actions incorrectly or performing extraneous and detrimental actions such as shutting down safety injection when it is not appropriate. See Section 6 of this report for more information on the inclusion of EOCs and some related good practices.

Finally, it should be recognized that iterations as well as refinement and review of the PRA model construction may (and often do) provide additional opportunities to identify human actions that need to be modeled.

5.1.4 Possible Impacts of Not Performing Good Practices and Additional Remarks

Failure to perform the above good practices could lead to an incomplete list of post-initiator human activities that if not completed correctly, may impact the outcome and frequencies of accident sequences addressed in the PRA. Therefore, this could result in potentially missing (i.e., not modeling) a risk-significant post-initiator human action that in turn, could make the model incomplete and/or inaccurate, potentially resulting in misinformation concerning the risk-significant plant features (including the important human actions).

Further, while not all the post-initiator actions will be important in the final assessment of risk, unlike the pre-initiator actions, it is difficult to predetermine (at this stage) a set of actions that do not have to be included as part of the identification process. Ways the operators interact with the plant equipment and affect the outcome of any accident sequence need to be assessed in order to determine their relative significance. Hence, the good practices herein are aimed at ensuring potentially risk-significant post-initiator actions (based on the procedures as well as the ways the procedures are interpreted and carried out) are identified at this stage of the analysis.

5.2 Modeling Specific HFEs Corresponding to Human Actions

5.2.1 Objective

The objective is to define how each specific post-initiator HFE is to be modeled in the PRA to accurately represent the failure of each action identified. This involves (1) the modeling of the HFEs as human-induced unavailabilities of functions, systems, or components consistent with the level of detail in the PRA accident sequences and system models, (2) possible grouping of responses into one HFE, and (3) ensuring that the modeling reflects certain plant- and accident sequence-specific considerations. The following provides good practices for modeling post-initiator HFEs while implementing RG 1.200 (Ref. 4) and the related requirements of the ASME Standard (Ref. 5).

5.2.2 Regulatory Guide 1.200 Position

The ASME Standard calls for the HFEs to be defined so that they represent the impact of not properly performing the required actions, consistent with the structure and level of detail of the accident sequences. There are multiple supporting requirements in the ASME Standard under high-level requirement HLR-HR-F that address the modeling level of detail for each HFE and how to complete the definition of each HFE. The regulatory guide states that the NRC staff has no objections to the ASME Standard requirements covering this activity.

5.2.3 Good Practices

5.2.3.1   Good Practice #1: Include HFEs for Needed Human Actions in the PRA Model

Define each specific post-initiator HFE to be modeled in the PRA as a basic event that describes the human failure of not properly performing the required actions and locate the HFE in the model such that it is linked to the unavailability of the affected component, train, system, or overall function depending on the effect(s) of the HFE (e.g., failure to manually depressurize using the safety relief valves, failure to manually scram, failure to align the backup train of service water). The following considerations should be used to properly define the post-initiator failure level in the PRA:

- whether the action is performed on a train, system, or component level, making it logical to define the HFE at the level of a whole train, system, or component

- whether the consequences of the failure and what would be affected by the failure are at the component, train, system, multiple systems, or entire function level

- whether multiple individual actions/responses such as at a system or component level (e.g., starting high-pressure injection and then subsequently opening a power-operated pressurizer relief valve) can be combined as a single post-initiator HFE affecting a higher level of equipment resolution such as at a system or a function level (e.g., initiating feed-and-bleed). This could be done as long as the following criteria are met:
  - ▸ The actions and effects are related.
  - ▸ The factors affecting the quantification of the single HFE will not be significantly different than those that would have been relevant for the individual actions (e.g., the same PSFs will be relevant) or the quantification result will be conservatively bounding compared to separately modeling and quantifying the individual actions.
  - ▸ There are no potential commonalities/dependencies with other post-initiator actions elsewhere in the model so that potential common failures among similar individual actions might be missed (see the discussion presented below).
  - ▸ The level of detail is already modeled in the PRA (e.g., train, system) for failures of the associated equipment (but note that this is a less important factor than those above).

As an example of how human responses may be grouped and modeled as one or more HFEs, consider the case in a boiling-water reactor (BWR) of a desired response to control reactivity in an anticipated transient without scram scenario. Failure to control reactivity could be defined as one HFE, or as several HFEs based on the subtasks involving inhibiting the automatic depressurization system, lowering reactor water level, and initiating the standby liquid control system.

For situations such as the above example, it is usually best to model separate HFEs if failures to perform the subtasks:

- have different effects
- may individually be impacted by very different PSFs (e.g., in-control room actions vs. local actions in a high-steam environment area, a subtask performed early in the scenario vs. another subtask performed much later in the scenario)
- involve an action that has a dependency with some other action to be modeled in the PRA (e.g., failure to trip two reactor coolant pumps followed by subsequent failure to trip the remaining reactor coolant pumps when conditions warrant)

An alternative is to model them all as one HFE and model the bounding consequence (such as the failure to control reactivity example cited above) as long as the most limiting PSFs are used (e.g., the shortest time that any of the subtasks must be performed, the most complex of the subtasks, etc.) and any subtask dependencies with other HFEs are identified, treated in the model, and properly quantified.

The failure mode defined by the HFE should be consistent with both the human failures and the equipment affected by these human failures (refer to the Good Practices in Section 5.1.3). The failures should sufficiently describe the HFE and its effect to ensure proper interpretation of the HFE in the model (e.g., fail to initiate feed-and-bleed within 5 minutes after the reactor pressure reaches 2,400 psig).

As an aid to ensure appropriate modeling, it is recommended practice that the post-initiator failure be placed in proximity, in the PRA model, to the component, train, system, or function affected by the human failure. In this way, a quick examination of the model can reveal the modeled effect of the human failure.

5.2.3.2 Good Practice #2: Define the HFEs Such that they are Plant- and Accident Sequence-Specific

Each of the modeled post-initiator HFEs should be defined such that they are plant- and accident sequence-specific, and the basic events representing them are labeled uniquely. In order for the action to occur, the operator must diagnose the need to take the action and then execute the action. While many PSFs are used to quantify the probability for failing to perform the action correctly because of either a diagnosis or implementation error (as discussed later under quantification), all of which should be evaluated based on plant- and accident sequence-specifics, the following requirements are particularly germane to a basic understanding of the HFE and should be met to complete the definition of each HFE:

- To the extent possible, the time by which the action needs to be performed (e.g., fail to initiate feed-and-bleed by 2 minutes after primary pressure reaches 2,400 psig), and the time necessary to diagnose the need for and to perform the action should be based on plant- and accident sequence-specific timing and nature of the complexity and/or subtasks involved in implementing the action. In other words, timing information should not be based on another plant's analysis or a general analysis for the "average" plant without proper justification, since the number and nature of the specific manipulations could be different, the plant thermal-hydraulic response could be different, the location for local actions may require different travel times, and some sequences require different response times for the same action, and so forth.

5-6

- Similar to the above, the availability and timing of plant- and accident sequence-specific cues (i.e., indications, alarms, visual observations, etc. and when they will be manifested) should be used as this timing information can differ plant-to-plant and sequence-to-sequence and will affect the likelihood and timing of diagnosing the need for the action.

- Plant-specific procedure and training guidance should be used based on the reviews under the Good Practices in Section 5.1.3.

- Where the action is performed (e.g., in the control room, locally in the auxiliary building) should be noted.

5.2.3.3　　Good Practice #3: Perform Talk-Throughs, Walkdowns, Field Observations, and Simulator Exercises (as Necessary) to Support the Modeling of Specific HFEs

To fully understand the nature of the action(s) (e.g., who performs it, what is done, how long does it take, whether there are special tools needed, whether there are environmental issues or special physical needs, whether there is a preferred order of use of systems to perform a specific function, etc.) and help define the HFEs and their context, additional reviews, talk-throughs, walkdowns, field observations, and simulator exercises are performed (as discussed in Good Practice #2 under Section 3.1.3.2, with more about the benefits of these techniques presented in Appendix B). In addition, the results of these activities may add to the list of actions and/or help interpret how procedural actions should be defined based on how they are actually carried out. Analysts should perform the following activities:

- Review training material and, where possible, perform talk-throughs or walkdowns of the actions with operations or training staff while going through the procedures to ensure consistency with training policies and teachings and to identify likely operator response tendencies for various conditions that may not be evident in the procedures. For example, operators may cite a reluctance to restart reactor coolant pumps in spite of the procedure direction based on their training and perceived adverse effects, or they may have a preference to use condensate as a BWR injection source before using lower pressure emergency core cooling system. These added "interpretations" of the procedures can help complete and/or clarify the identified actions and ensure that later modeling and quantification of the actions will reflect the "as-operated" plant.

- Observe simulator exercises of accident scenarios since these can provide valuable insights with regard to how the actions are actually carried out, by whom, and particularly how procedure steps are interpreted by plant crews, especially where the procedure is ambiguous or leaves room for flexibility in the crew response. For example, through simulation it may be observed that a "single action" in the procedure (e.g., align recirculation) is actually carried out by a series of numerous and sequential individual actions (e.g., involving the use of many handswitches in a certain sequence). Again these observed "interpretations" of the procedures can help complete and/or clarify the identified actions and ensure that later modeling and quantification of the actions will reflect the "as-operated" plant.

5.2.4 Possible Impacts of Not Performing Good Practices and Additional Remarks

Failure to perform the above good practices could lead to improper modeling of the HFE and thus a misrepresentation of its effects on the plant equipment and sequence outcome. Depending on the degree of failure to follow the above good practices, the risk effect of failing to take the proper action could be inappropriately overemphasized (such as improperly linking the action to more equipment than is actually affected or using too conservative thermal-hydraulic information for the relevant timing associated with the action), underestimated (such as if combining many actions into too broad a single HFE without properly bounding the consequences of the failure or not accounting for some plant- and accident sequence-specific negative PSFs as confirmed by simulations), or even missed entirely (such as not reviewing all appropriate procedures). This can result in inaccuracies in the PRA results and particularly incorrect assessments of the importance of post-initiator human failures. The precise definition of the post-initiator basic events and their placement in the model (from both a logic and failure mode standpoint) ultimately define how the model addresses the effects of the human failures. This needs to be done accurately if the model is going to logically represent the real effects of each human failure and if the corresponding HFE is going to be correctly quantified (as discussed later). This accuracy is best obtained if plant- and accident sequence-specific information is used.

Nonetheless, we note that not using plant- and accident sequence-specific thermal-hydraulic information for timing may or may not be critical based on the relevancy and thus appropriateness of the non-specific (i.e., "general") timing information that is used. It is better to use plant- and accident-specific information, though it is recognized that in some areas (e.g., containment response for LERF), from a practical standpoint, modified "general" information may be all that is readily available. Further, as long as the timing considerations used are reasonable and accurate to within the resolution of the HRA quantification tool to be used, differences between plant- and accident-specific versus more "general" timing considerations may not be a significant issue. Analysts should ensure that if non-specific timing information is used, the timing is reasonable for the plant and accident sequence being analyzed.

5.3   Quantifying the Corresponding HEPs for Post-Initiator HFEs

5.3.1   Objective

The objective is to address how the HEPs for the modeled HFEs from the previous analysis activity are to be quantified. This section provides good practices on an attribute or criteria level and does not endorse a specific tool or technique. Ultimately, these probabilities (along with the other equipment failure and pre-initiator HEPs) and initiating event frequencies are all combined to determine such risk metrics as CDF, LERF, ΔCDF, ΔLERF, etc., as addressed in RG 1.174 (Ref. 3). The following provides good practices for quantifying post-initiator HEPs while implementing RG 1.200 (Ref. 4) and the related requirements of the ASME Standard (Ref. 5).

5.3.2 Regulatory Guide 1.200 Position

The ASME Standard requires that a well-defined and self-consistent process be used to quantify the post-initiator HEPs. There are multiple supporting requirements in the Standard under high-level requirement HLR-HR-G that address many factors associated with quantifying the HEPs. These include, when conservative vs. detailed estimates are appropriate, consideration of both diagnosis and execution failures, PSFs considered in the evaluations, consideration of dependencies among HFEs, uncertainty, and reasonableness of the HRA results. The regulatory guide states that the NRC staff has only one clarification to the ASME Standard requirements covering this activity, and that is to ensure that the availability of staff resources to carry out all of the desired actions is considered. This has been addressed among the PSFs covered in the good practices.

5.3.3 Good Practices

5.3.3.1 Good Practice #1: Address Both Diagnosis and Response Execution Failures

Whether using conservative or detailed estimation of the post-initiator HEPs, the evaluation should include both diagnosis and execution failures. For example, incorrectly interpreting a cue or not seeing a cue and thus not performing the action can be one mode of failure. Alternatively, the operator can intend to take the action based on the proper and recognized cues, but still otherwise fail to take the action or perform it correctly. Both need to be part of the HEP evaluations. However, the qualitative HRA analysis may indicate that one of these failure modes predominates the other in such a way that the effect of only one failure mode needs to be quantified, but this should be justified.

5.3.3.2 Good Practice #2: Use Screening Values During the Initial Quantification of the Post-Initiator HFEs

The use of conservative HEP estimates (screening values) is usually desirable during the early stages of PRA development and quantification. This is acceptable (and almost necessary since not all the potential dependencies among human events can be anticipated) provided (1) it is clear that the individual values used are overestimations of the probabilities that would be developed if detailed assessments were to be performed, **and** (2) dependencies among multiple HFEs appearing in an accident sequence are conservatively accounted for. These screening values should be set so as to make the PRA quantification process more efficient (by not having to perform detailed analysis on every HFE), but not so low that subsequent detailed analysis would actually result in higher HEPs. The screening estimates should consider both the individual events and the potential for dependencies across multiple HFEs in a given accident sequence (scenario). To meet these conditions, the following conditions are recommended:

- No conservative HEP value assigned to an individual post-initiator HFE should be lower than the worse case anticipated detailed value and generally not lower than 0.1 (which is typical of high post-initiator values in PRAs, but 0.5 is also often appropriate and sometimes used).

- Multiple HFEs in the same sequence should not have a joint probability value lower than the worse case anticipated detailed joint probability value and generally not lower than 0.05 (accounts for a 0.5 high dependency factor) at this stage.

5.3.3.3 Good Practice #3: Perform Detailed Assessments of Significant Post-Initiator HFEs

As needed for the issue being addressed to produce a realistic assessment of risk, detailed assessments of the significant HFE contributors should be performed. [See Table A-1 in RG 1.200 (Ref. 4) for a definition of "significant contributor."] The PRA analyst can define the significant contributors by use of typical PRA criteria [not addressed here, but see Section 2.3 of NUREG-1764 (Ref. 8)], such as importance measure thresholds as well as other qualitative and quantitative considerations. Note that performing the detailed assessments may be an iterative process, since beginning the process may lead to the need to perform other detailed assessments as HEPs are changed in the model. Conservative estimates, or screening values should not be used in place of detailed analysis. While the use of conservative values may, at first, seem to be a "safe" analysis process, it can have negative impacts. More conservative values can focus the risk on the wrong human actions or related accident sequences and equipment failures because of the intentionally high HEPs. Such incorrect conclusions need to be avoided by ensuring a sufficient set of more realistic, detailed HEPs are included in the model. In fact, non-significant HFEs should also be assessed in detail if detection of *all* weaknesses in plant design or practices is desirable given the application.

5.3.3.4 Good Practice #4: Revisit the Use of Post-Initiator Screening Values vs. Detailed Assessments for Special Applications

For a specific PRA application and depending on the issue being addressed (e.g., examination of a specific procedure change), revisit the use of conservative (screening) vs. detailed (assessed) HEPs to ensure issue-relevant human actions have not been prematurely deleted from the PRA or there is an inappropriate use of conservative vs. detailed values to properly assess the issue and the corresponding risk.

5.3.3.5 Good Practice #5: Account for Plant- and Activity-Specific PSFs in the Detailed Assessments of Post-Initiator HEPs

As "good practice," the following table of PSFs (Table 5-1) for both in-control room and local (ex-control room) actions should be treated in the evaluation of each HEP per the table guidance. The guidance should fit most cases, but it should be recognized that for specific actions, some of the factors may not apply. Also, there may be HFEs and contexts for which some PSFs are so important, the others do not matter (e.g., time available is so short, the action almost assuredly cannot be done regardless of the other factors). Further, if a specific situation warrants treatment of unique factors that are not, and cannot be addressed by the list of factors in the table, identification of other PSFs should complement the list. Note that the guidance provided in Table 5-1 on the conditions under which certain PSFs may be particularly relevant is meant solely to serve as a general guide. While the goal of the information in that column is to point out situations likely to make certain PSFs important, the table is meant to be illustrative and certain PSFs may also be important under other conditions. Thus, a given PSF should not be ignored simply because the specific conditions in the table are not present. Consideration of the impact of the factors on the HEPs should be as plant- and accident sequence-specific as necessary to address the issue. In addition, such influences should be confirmed, where useful, by such techniques as talk-throughs, walkdowns, field observations, simulations, and examination of past events in order to be realistic. Appendix B provides more specific guidance and discussion of the PSFs presented in the table.

It should be noted that there are aleatory aspects of some or all of the PSFs listed below and not all of these aleatory influences will need to be considered for all applications. For example, for a license amendment, the more "systematic" PSFs (e.g., procedures, training) or the more systematic aspects of the PSFs (e.g., for the team/crew dynamics PSF, the plant practice for the division of control room responsibilities would tend to be systematic) may be the more important ones. On the other hand, when performing assessments of specific events, some of the more aleatory factors or aspects (e.g., crew aggressiveness, failed instrumentation, etc.) may be directly relevant. Nonetheless, for any application, there may be aleatory factors that could be very important for given scenarios and analysts should at least consider whether various aleatory influences could be important enough in the context of the event to be considered directly. For example, if the timing of the scenario is such that only aggressive crews will be able to reach the relevant step in the procedure soon enough, and it is determined that the plant operating crews vary significantly in terms of their aggressiveness in implementing the procedures, then direct modeling of this variable may be appropriate. At a minimum, such aleatory factors should be considered in determining the mean HEP value and in consideration of the uncertainty.

The analysis should ensure that the factors seemingly most relevant to the HFE (either as positive or negative influences) and having the most impact on the HEP, have been considered quantitatively. Furthermore, the quality of the HFE evaluations is improved when the impacts of relevant factors have been determined from collected information (e.g., talk-throughs, walkdowns, field observations, simulations) rather than simple judgments.

Table 5-1 Post-Initiator PSFs To Be Considered for Both Control Room and Local (Ex-Control Room) Actions

Post-Initiator PSFs To Consider (Relative to Each Scenario and HFE)	Conditions When Particularly Relevant	
	Control Room Actions	Local Actions*
Applicability and suitability of training and experience	Always	Always
Suitability of relevant procedures and administrative controls	Always	Always
Availability and clarity of instrumentation (cues to take actions as well as confirm expected plant response)	Always	When time is short
Time available and time required to complete the action, including the impact of concurrent and competing activities	Always	Always
Complexity of required diagnosis and response. In addition to the usual aspects of complexity, special sequencing, organization, and coordination can also be contributors to complexity.	Unfamiliar Situation	Unfamiliar Situation
Workload, time pressure, stress	Crew is aware of time constraints	Crew is aware of time constraints

Post-Initiator PSFs To Consider (Relative to Each Scenario and HFE)	Conditions When Particularly Relevant	
	Control Room Actions	Local Actions*
Team/crew dynamics and crew characteristics [degree of independence among individuals, operator attitudes/biases/rules, use of status checks, approach for implementing procedures, (e.g., aggressive vs. slow and methodical)]. Note: Observation of simulator exercises and discussions with operating crews and trainers are particularly important to obtaining this type of information. Weaknesses and strengths in organizational attitudes and rules as well as in administrative guidance may bear on aspects of crew behavior and should be considered.	Always	When the timing and the appropriateness of the directions from the control room (CR) is critical or could be affected by the scenario, or when the subsequent carrying out of the local [ex-CR action(s)] involves teamwork.
Available staffing and resources	If typical CR staff is expected to be decreased or impacted so that others must perform more than their typical tasks. (This is not usually an issue, but a fire scenario might be an exception.)	Particularly when many or complex actions need to occur concurrently or in a short time, and staffing needs may be stretched. Time of day may be a contributing factor to the availability of staff.
Ergonomic quality of human-system interface (HSI)	If could be problematic, or not easily accessed or used (not usually an issue, but consider, for instance, the need to use backboards, deal with common workarounds)	If could be problematic (e.g., poor labeling) or not easily accessed or used
Environment in which the action needs to be performed	Potentially adverse or threatening situations such as fire, flood, seismic, loss of ventilation (not usually an issue)	Potentially adverse situations such as high-radiation, high-temperature, high-humidity, smoke, toxic gas, noise, poor lighting, weather, flooding, seismic

Post-Initiator PSFs To Consider (Relative to Each Scenario and HFE)	Conditions When Particularly Relevant	
	Control Room Actions	Local Actions*
Accessability and operability of equipment to be manipulated	If operability could be problematic, or not easily accessed or used, such as the need to use backboards, or when indications/controls could be affected by the initiating event or other failures [e.g., loss of direct current (DC)]	If operability could be problematic (e.g., rarely used), or not easily accessed, such as when the equipment could be affected by the initiating event (e.g., fire, flood, loss of power)
The need for special tools (keys, ladders, hoses, clothing such as to enter a radiation area)	Not usually an issue, but consider, for instance, when accessability of keys for keylock switches could be important	For situations where other than simple switch or similar type operations are necessary, or when needed to be able to access the equipment
Communications (strategy and coordination) as well as whether one can be easily heard	Not usually an issue. Simply ensure that communication strategy allows crisp direction and proper feedback; otherwise only in special situations such as needing to communicate while wearing a self-contained breathing apparatus (SCBA)	For situations where communication among crew members (locally and/or with CR) are likely to be needed and there could be a threat such as too much noise, failure of the communication equipment, availability and location issues associated with the communication equipment.
Special fitness needs	Typically not an issue	For special situations expected to involve the use of heavy or awkward tools/equipment, carrying hoses, climbing, etc.

Post-Initiator PSFs To Consider (Relative to Each Scenario and HFE)	Conditions When Particularly Relevant	
	Control Room Actions	Local Actions*
Consideration of "realistic" accident sequence diversions and deviations (e.g., extraneous alarms, failed instruments, outside discussions, sequence evolution not exactly like that trained on). Note: This item is essentially addressing aleatory factors that could have important effects on performance. While analysts may choose to explicitly address these factors only when a more detailed investigation of the scenario is warranted or when they are explicitly part of the question being asked, it should be recognized that in some cases they could have strong effects. If they are not addressed explicitly in the analysis, it is suggested that their potential impacts be considered in assessing the HEP values. (See the related discussion in Section 5.3.3.7.)	When reasonable variations in plant conditions for the scenario (even if of low probability) have the potential to confuse operating crews.	Rarely important, but to the extent that these aspects could affect the timing, specific directions, or successful performance of the local action(s), they should be considered.
* Note that the decision to take local actions typically takes place in the control room (e.g., the shift supervisor decides to dispatch an operator to a location) and, thus, certain control room related PSFs will likely apply to the decision process to take local actions.		

5.3.3.6 Good Practice #6: Account for Dependencies Among Post-Initiator HFEs

Dependencies among the post-initiator HFEs and hence the corresponding HEPs in an accident sequence should be quantitatively accounted for in the PRA model by virtue of the joint probability used for the HEPs. This is to account for the evaluation of each sequence holistically, considering the performance of the operators throughout the sequence response and recognizing that early operator successes or failures can influence later operator judgments and subsequent actions. This is particularly important so that combined probabilities that are overly optimistic are not inadvertently assigned, potentially resulting in the inappropriate decrease in the risk-significance of human actions and related accident sequences and equipment failures. In the extreme, this could result in the inappropriate screening out of accident sequences from the model because the combined probability of occurrence of the events making up an accident sequence drops below a threshold value used in the PRA to drop sequences from the final risk results.

In analyzing for possible dependencies among the HFEs in an accident sequence, and considering dependency factors commonly treated in various current HRA techniques along with the few examples provided in the ASME Standard (Ref. 5), the following provides guidance on the search for links among the actions represented in different HFEs:

*   The same crew member(s) is responsible for the actions.

*   The actions take place relatively close in time such that a crew "mindset" or interpretation of the situation might carry over from one event to the next.

- The procedures and cues used along with the plant conditions related to performing the actions are identical (or nearly so) or related, and the applicable steps in the procedures have few or no other steps in between the applicable steps.

- Similar PSFs apply to actions.

- How the actions are performed is similar and they are performed in or near the same location.

- There is reason to believe that the interpretation of the need for one action might bear on the crew's decision regarding another action, i.e., the basis for one decision in a scenario may influence another decision later in the scenario.

The more the above commonalities and similarities exist, the greater the potential for dependence among the HFEs (i.e., if the first action is not performed correctly, there is a higher likelihood that the second, third... action(s) will also not be performed correctly; and conversely if the action(s) are successful). For example, if nearly all or all of the above characteristics exist, very high or complete dependence should generally be assumed. If only one or two of the above characteristics exist, then analysts will need to evaluate the likely strength of their effect and the degree of dependence that should be assumed and addressed in quantification.

The resulting joint probability of the HEPs in an accident sequence should be such that it is in line with the above characteristics and the following guidance, unless otherwise justified:

- The total combined probability of all the HFEs in the same accident sequence/cut set should not be less than a justified value. It is suggested that the value not be below ~0.00001 since it is typically hard to defend that other dependent failure modes that are not usually treated (e.g., random events such as even a heart attack) cannot occur. Depending on the independent HFE values, the combined probability may need to be higher.

- To the extent the joint HEPs are looked at separately, but a previous human action in the sequence has failed, expectations as to typically assigned dependency factors based on available tools/techniques are provided here:
  - A factor of 3 to 10 higher than what would have been the independent HEP value for the subsequent action(s) exists for low to moderate dependence
  - 0.1 up to 0.5 is the resulting probability value used for the subsequent HEP(s) for high dependence
  - ≥ 0.5 exists for the subsequent HEP(s) for very high or 1.0 for complete dependence.

5.3.3.7 Good Practice #7: Assess the Uncertainty in Mean HEP Values

Mean values for each HEP (excluding conservative HEPs) should be obtained and an assessment of the uncertainty in the HEP values should be performed at least for the significant HEPs to the extent that these uncertainties need to be understood and addressed in order to make appropriate risk-related decisions. Assessments of uncertainty are typically performed in the following ways:

- developing uncertainty distributions for the HEPs

- propagating them through the quantitative analysis of the entire PRA, such as by a Monte Carlo technique

- performing sensitivity analyses that demonstrate the effects on the risk results for extreme estimates in the HEPs based on at least the expected uncertainty range

Note, in some cases, it may be sufficient to address the uncertainties only with qualitative arguments without the need to specifically quantify them (e.g., justifying why the HEP cannot be very uncertain or why a change in the HEP has little relevancy to the risk-related decision to be made).

Generally, however, a more thorough treatment of uncertainty is needed. While the guidance provided below describes good practice for developing uncertainty distributions, in most cases, it may go beyond what has been done in even recent PRA applications. Nonetheless, analysts should consider such approaches to the extent necessary to ensure that their application is as realistic as necessary. In other words, they need to examine the question being addressed and ask whether a more thorough treatment of uncertainty would be beneficial.

Thus, it is recommended that in assessing the uncertainties and particularly when assigning specific uncertainty distributions, the uncertainties should include the following:

- those epistemic uncertainties existing because of lack of knowledge of the true expected performance of the human for a given context and associated set of PSFs (i.e., those factors for which we do not have sufficient knowledge or understanding as to the "correct" HEP, such as how time of day affects the biorhythm and, hence, performance of operators)

- consideration of the combined effect of the relevant aleatory (i.e., random) factors *to the extent they are not specifically modeled in the PRA* and to the extent that they could significantly alter the context and PSF evaluations for the HFE, and thereby the overall HEP estimate

Concerning the latter, it is best to specifically model the aleatory factors in the PRA (i.e., those factors that are random and could significantly affect operator performance, for example, whether or not other nuisance alarms or equipment failures may coexist with the more important failures in the sequence, whether a critical equipment failure occurs early in the sequence or late in the sequence, etc.). However, this is often impractical and is typically not done as it would make the PRA model excessively large and unwieldy. Thus, in assigning the mean HEP and uncertainty distribution, analysts should reflect an additional contribution from random factors associated with the plant condition or overall action context. This can be done by considering the relevant aleatory (i.e., random) factors, their likelihoods of occurrence, and their effects on the HEP estimate.

For example, suppose for an accident sequence, it is judged that the human performance will be significantly affected by the number of "nuisance and extraneous failures," as opposed to when no or few nuisance/extraneous failures exist (and yet these two plant "states" are not explicitly defined by the PRA model). Further, based on the analyst considering how the HEP is affected, a value of P_o would be estimated for when no or few nuisance/extraneous failures exist and a value of P_1 would be estimated for when many do exist, and the difference between P_o and P_1 is significant (e.g., factor of 10). It is also judged that many nuisance/extraneous failures will occur about 50% of the time based on past experience. The resulting combined mean HEP value is $0.5P_o + 0.5P_1$ considering this random factor. The overall uncertainty about the combined mean HEP value should reflect the weighted epistemic uncertainties in P_o and P_1 (such as by a convolution approach, via an approximation, or other techniques). While it is not expected that such a detailed evaluation will be done for every random situation or every HEP, the mean and uncertainty estimates for the most significant HEPs should account for any such perceived important aleatory factors that have not otherwise been accounted for (i.e., if the factors, considering their likelihoods and effects on the HEP, are anticipated to have a significant impact on the resulting overall HEP).

Whatever uncertainty distributions are used, the shape of the distributions (log-normal, beta, etc.) is typically unimportant to the overall risk results (i.e., the PRA results are usually not sensitive to specific distributions). Further, typical uncertainties include values for the HEP that represent a factor of 10 to 100 or even more between the lower bound value and the upper bound value that encompass the mean value. However, it should be noted that some distributions (e.g., log-normal) can give probabilities greater than 1.0 for HEPs that are relatively high.

5.3.3.8 Good Practice #8: Evaluate the Reasonableness of the HEPs Assigned Using Detailed Assessments

The HEPs for post-initiator HFEs (excluding the HFEs assigned conservative/screening HEPs) should be reasonable from two standpoints:

- first and foremost, relative to each other (i.e., the probabilistic ranking of the failures when compared one to another)
- in absolute terms (i.e., each HEP value), given the context and combination of positive and negative PSFs and their relative strengths

This reasonableness should be checked based on consideration of actual plant experience and history, against other evaluations (such as for similar actions with similar context and PSFs at other plants), and the qualitative understanding of the actions and the relevant contexts and PSFs under which the actions are performed. It is suggested that a rank-ordered list of the post-initiator HFEs by probability be used as an aid for checking reasonableness. As part of such a list, it is particularly worthwhile to compare "like" HFEs for different sequences such as failure to manually depressurize in a BWR when all high-pressure injection is lost during a small loss-of-coolant accident (LOCA), compared to the same action during a simple transient. For example, simple, procedure-guided actions with easily recognized cues and plenty of time to perform the actions, should have lower HEPs than complex, memorized, short time available type actions, all other factors being the same. Typical expectations of most post-initiator HEPs are in the range of 0.1 to 0.0001 and depend particularly on the relevant contextual factors and proper consideration of dependencies as discussed under many of the good practices covered above. Helpful checks include the following:

- For an HFE, is there one or two dominant PSFs that exist or is the cumulative effect of the relevant PSFs such that they are either so negative or so positive that a "sanity check" would suggest a high HEP (e.g., 0.1) or a low HEP (e.g., 0.0001), respectively? For example, if the procedures and training fit the scenario well, the cues will be clear, there is plenty of time available, and the actions are simple and familiar, then an HEP of 0.001 to 0.0001 would be reasonable. Alternatively, if there is a reasonable expectation that the scenario could be confusing, it is not frequently trained on, there may be some hesitancy associated with taking the action, and only an aggressive crew would be likely to diagnose and complete the action in the time available, then an HEP of 0.5 or even 1.0 would not be unreasonable. Accordingly, this very high- or low-probability HFE should be one of the higher- or lower-probability HFEs, relative to the other HFEs in the model.

- Are there seemingly balanced combinations of both positive and negative factors, or are there weak to neutral factor effects? If so, this is likely to lead to in-between values for the HEPs (e.g., ~0.01) placing these HFEs (relative to others) "in the middle."

- Do the individual HEPs and the relative ranking of the HFEs seem consistent with actual or simulated experience? For example, if it is known that operators "have trouble with" a specific action(s) in simulations or practiced events, and yet the assigned HEP is very low (e.g., 0.001 or lower), this may be a reason to question and revisit the assigned HEP.

- Do other similar plant and action analyses support the HEP evaluation? Recognize, however, that there may be valid reasons why differences may exist and thus this check is not likely to be as helpful as the others above.

5.3.4 Possible Impacts of Not Performing Good Practices and Additional Remarks

Failure to quantify the post-initiator HEPs as realistically as possible using the good practices articulated above (except for where higher screening estimates are used, purposely and appropriately), could result in improper HEPs and thus inaccuracies in the PRA results and particularly incorrect assessments of the importance of pre-initiator human failures. The risk effect of the human failure could be overemphasized [such as if the human failure is estimated with too high (pessimistic) an HEP or a high screening estimate is used where a more realistic detailed estimate is appropriate], or underestimated [such as if the human failure is estimated with too low (optimistic) an HEP or a dependency among failures is not accounted for resulting in too low a joint probability for multiple human failures]. Besides these concerns about inaccuracies in the HEP quantification and thus whether the HEPs "make sense," as well as the resulting potential misinformation about the significant risk contributors if quantification is not done well, the following related observations are noted:

- Use of conservative values is a useful and most often necessary part of HRA so as to avoid the expenditure of resources on unimportant human events and accident sequences. The above guidance is aimed at allowing some conservative values without inadvertently and inappropriately allowing the analytical phenomenon of, for instance, multiplying four human events in the same sequence each at a conservative estimate of 0.1 to yield a 0.0001 combined probability, without checking for dependencies among the human events. In such cases, some HFEs and combinations of events, or even whole accident sequences, may inappropriately entirely screen out of the PRA model because the accident sequence frequency drops below a model threshold. Hence, some of the significant individual or combination contributors may be missed. This is why the conservative estimates both individually and for combined events should not be too low. Further, if conservative values are left permanently assigned to some HFEs that should be assessed with more detail to obtain a more realistic assessment of risk (supposedly lowering the probability), the risk-significance of these HFEs and related equipment failures are likely to be overemphasized at the expense of improperly lessening the relative importance of other events and failures.

- It is important to be sure that dependencies among the various modeled HFEs including those with conservative values, have been investigated. Treating HFEs, whether with conservative values or based on more detailed analysis, as independent actions without checking for dependencies (thereby being able to multiply the individual HEPs) can inappropriately lessen the risk-significance of those HFEs and related equipment failures in accident sequences. This can cause the inappropriate dropping out of accident sequences because the sequences quantitatively drop below a model threshold value as discussed above under screening. Proper consideration of the dependencies among the human actions in the model is necessary to reach the best possible evaluation of both the relative and absolute importance of the human events and related accident sequence equipment failures.

- The use of mean values and addressing uncertainties are a part of the guidance in RG 1.174 (Ref. 3) and, to the extent addressed therein, the HRA quantification needs to be consistent with that guidance when making risk-informed decisions. The estimates should reflect, to the extent possible, the as-built and as-operated conditions as addressed in the plant- and activity-specific PSFs.

- There can be a tendency to use an existing PRA model to address issues such as changes to the plant, without spending the appropriate time to revisit some of the underlying assumptions and modeling choices made to create the original PRA. However, such a review should be done to see if these assumptions and choices still apply for the issue being addressed. For instance, some post-initiator HFEs may be quantified in the original model using conservative estimates and detailed failure probabilities that may not be appropriate for the new issue being addressed.
As an example, where high conservative HEPs may have been acceptable for purposes of the original PRA, these may overestimate the contribution of these HFEs for the issue being addressed. Further, the relative risk contribution of equipment and associated accident sequences with which the HFEs appear, may be artificially too high (and, therefore, other events too low) because of the conservative values. Hence, it is good practice to revisit the use of conservative estimates and detailed HFE probabilities in order to appropriately address the issue.

5.4   Adding Recovery Actions to the PRA

5.4.1   Objective

The objective is to address what recovery actions can be credited in the post-initiator HRA and the requirements that should be met before crediting recovery actions. Adding recovery actions is common practice in PRA and accounts for other reasonable actions the operators might take to avoid severe core damage and/or a large early release that are not already specifically modeled. For example, in the PRA modeling of an accident sequence involving loss of all injection, it would be logical and common to credit the operators attempting to locally align an independent firewater system for injection. The failure to successfully perform such actions would subsequently be added to the accident sequence model thereby crediting the actions and further lowering the overall accident sequence frequency because it takes additional failures of these actions before the core is actually damaged. The following provides good practices for crediting post-initiator recovery actions while implementing RG 1.200 (Ref. 4) and the related requirements of the ASME Standard (Ref. 5).

5.4.2   Regulatory Guide 1.200 Position

The ASME Standard requires that recovery actions be modeled only if it has been demonstrated that the actions are plausible and feasible for those sequences to which they are applied. There are multiple supporting requirements in the Standard under high-level requirement HLR-HR-H that address what recovery actions can be credited as well as the need to consider dependencies among the HFEs and any recovery actions that are credited. The regulatory guide states that the NRC staff has only one clarification to the ASME Standard requirements covering this activity, and that concerns factors that need to be considered when accounting for dependencies between the recovery HFE and other HFEs in the sequence. This has been addressed in the good practices.

5.4.3   Good Practices

5.4.3.1   Good Practice #1: Define Appropriate Recovery Actions

Based on the failed functions, systems, or components, identify recovery actions[2] to be credited that are not already included in the PRA (e.g., aligning another backup system not already accounted for) and that are appropriate to be tried by the crew to restore the failure. The following should be considered in defining appropriate recovery actions:

*   the failure to be recovered

*   whether the cues will be clear and provided in time to indicate the need for a recovery action(s), and the failure that needs to be recovered

---

[2]   Note that the definition of a recovery action and its distinction from a repair action has been adopted from RG 1.200 (Ref. 4). *Recovery action* is defined as: a PRA modeling term representing restoration of the function caused by a failed SSC, by bypassing the failure. Such a recovery can be modeled using HRA techniques regardless of the cause of the failure. *Repair* is defined as a general term describing restoration of a failed SSC by correcting the failure and returning the failed SSC to operability. HRA techniques cannot be used since the method of repair is not known without knowing the specific causes.

- the most logical recovery action(s) for the failure, based on the cues that will be provided

- the recovery is not a repair action (e.g., the replacement of a motor on a valve so that it can be operated)

- whether sufficient time is available for the recovery action(s) to be diagnosed and implemented to avoid the undesired outcome

- whether sufficient crew resources exist to perform the recovery(ies)

- whether there is procedure guidance to perform the recovery(ies)

- whether the crew has trained on the recovery action(s) including the quality and frequency of the training

- whether the equipment needed to perform the recovery(ies) is accessible and in a non-threatening environment (e.g., extreme radiation)

- whether the equipment needed to perform the recovery(ies) is available in the context of other failures and the initiator for the sequence/cut set

In addressing the above issues and assessing which recovery action, or actions, to credit in the PRA, for post-initiator HFEs all the good practices provided earlier in Sections 5.1, 5.2, and 5.3 apply (i.e., the failure to recover is merely another HFE, like all of the other post-initiator HFEs). In general, no recovery should be credited where any of the above considerations are not met (e.g., there is not sufficient time, there are no cues that there is a problem, there are not sufficient resources, there is no procedure or training). It may be possible to justify exceptions in unique situations, such as a procedure is not needed because the recovery is a skill-of-the-craft, non-complex, and easily performed; or the specific failure mode of the equipment is known for the sequence (this is usually not the case at the typical level of detail in a PRA) and so "repair" of the failure can be credited because it can be easily and quickly diagnosed and implemented. Any exceptions should be documented as to the appropriateness of the recovery action.

When considering multiple recoveries (i.e., how many recoveries to be credited in one accident sequence/cut set), the above considerations apply to all the recoveries. The analyst should also consider that one recovery may be tried (perhaps even multiple times) and then the second recovery may be tried but with even less time and resources available because of the attempts on the first recovery. Hence, the failure probability of the second recovery should be based on more pessimistic characteristics (e.g., less time available, less resources) than if such a possibility is not considered.

5.4.3.2 *Good Practice #2: Account for Dependencies*

As stated above, all the good practices provided earlier in Sections 5.1, 5.2, and 5.3 apply. From these good practices, particular attention should be paid to accounting for dependencies among the HFEs including the credited recovery actions. More specifically, dependencies should be assessed:

- among multiple recoveries in the accident sequence/cut set being evaluated

- between each recovery and the other HFEs in the sequence/cut set being evaluated

As part of this effort, the analyst should give proper consideration to the difficulties people often have in overcoming an initial mindset, despite new evidence. For a real world example, consider how long the power-operated relief valve (PORV) path remained open in the Three Mile Island accident, despite new cues of the problem, different personnel arriving, etc. For this and similar reasons, the assessing of no dependence needs to be adequately justified to ensure the quantified credit for the recovery action(s) is not unduly optimistic.

5.4.3.3 Good Practice #3: Quantify the Probability of Failing to Perform the Recovery(ies)

Quantify the probability of failing to perform the recovery(ies) in either of the following ways:

- using representative data that can be shown to be appropriate for the recovery event(s) (e.g., using data that exists for typical times to reopen the main steam isolation valves and restore main feedwater)

- using the HRA method/tool(s) used for the other HFEs (i.e., using an analytical/modeling approach)

In performing the quantification, one should ensure that all the good practices under Section 5.3 are followed (for each individual recovery as well as for multiple/joint recovery credit). In addition, if using data, ensure the data is applicable for the plant/sequence context or that the data is modified accordingly. For example, a plant may use available experience data for the probability of failing to align a firewater system for injection but the experience data is based on designs for which all the actions can be taken from the main control room whereas for this plant, the actions have to be performed locally.

5.4.4 Possible Impacts of Not Performing Good Practices and Additional Remarks

Failure to follow the above good practices is likely to lead to recovery credit that is applied too optimistically; that is, the failure to recover is assigned too low a probability. Hence, an underestimate of the failure to recover is applied to the PRA accident sequence/cut set, making the affected sequence/cut set artificially too low in risk-significance. This can subsequently affect the ranking of the important sequences, equipment failures, and human actions, potentially leading to false conclusions as to the significant risk contributors.

6. ERRORS OF COMMISSION

Explicit modeling of errors of commission[3] has generally been beyond current PRA practice and is not explicitly addressed in RG 1.200 (Ref. 4) or ASME Standard (Ref. 5) HRA requirements. Many practitioners believe that the number of potential operator actions that could result in an EOC is unmanageably large. That is, they believe that a large number of potential actions would be averse to safe shutdown (e.g., fail or make unavailable equipment/functions relevant to mitigating the scenario, or otherwise exacerbate the scenario such as by opening a PORV and causing an unwanted LOCA), even for what may appear to be justifiable reasons. Errors of omission[4] are typically modeled in PRAs because they are easily defined and limited by the requirements of operating procedures (especially emergency operating procedures). At best, PRAs have handled EOCs implicitly (e.g., as part of a base HEP) without a systematic or adequate search for this type of error.

However, the need to consider EOCs has long been recognized (e.g., NUREG-1050, Ref. 31), and work in the area over the years [e.g., ATHEANA (Ref. 11), CESA (Ref. 15), Julius et al. (Ref. 16), Macwan and Mosleh (Ref. 17), Vuorio and Vaurio (Ref. 18), and Wakefield (Ref. 19)] has made advances in the ability to identify EOCs without the need to perform an exhaustive search. One of the lessons learned from the development and application of ATHEANA (Ref. 11) is that the effort needed to identify EOCs can be substantially reduced by focusing the search on identifying systematic vulnerabilities in plant operations associated with plant critical functions.

Furthermore, reviews of real-world serious accidents (e.g., ATHEANA, Ref. 11) have shown that EOCs can and do occur and even if in many instances the probabilities of the conditions that lead to them may be small, the likelihood of the actions given those conditions can be very high. The Three Mile Island Unit 2 accident, in which operators shut down all reactor cooling water injection because of the mistaken belief that the system was full of water, is a classic example of the potential importance of EOCs.

Finally, the NRC staff believes that plant changes do have the potential to create the opportunity for such failure modes and it has been recognized that many currently planned operator actions, e.g., manual operator actions in response to fires, could create problems. Thus, the NRC's position is that, for certain applications, it is a good practice to consider potential EOCs.

---

[3]   Error of Commission (EOC): A *human failure event* resulting from an overt, unsafe action, that, when taken, leads to a change in plant configuration with the consequence of a degraded plant state. Examples include terminating running safety-injection pumps, closing valves, and blocking automatic initiation signals.

[4]   Error of Omission (EOO): A *human failure event* resulting from a failure to take a required action, that leads to an unchanged or inappropriately changed plant configuration with the consequence of a degraded plant state. Examples include failures to initiate standby liquid control system, start auxiliary feedwater equipment, and failure to isolate a faulted steam generator.

6.1   Including and Modeling Errors of Commission

6.1.1   Objective

The objective is to describe the conditions under which EOCs should be considered for inclusion in the HRA modeling.

6.1.2   Regulatory Guide 1.200 Position

The regulatory guide does not provide any specific treatment of EOCs.

6.1.3   Good Practices

6.1.3.1   Good Practice #1: Address EOCs in Future HRAs/PRAs (Recommendation)

Given the recent advances in the ability to address EOCs and the potential for regulatory requirements to make the need to address EOCs more important, it is recommended that future HRA/PRAs attempt to identify and model potentially important EOCs. Sources available to support this process are listed above. It is suggested that multiple sources be consulted to assist this process.

To the extent any EOCs are modeled, all the guidance in this report has been written with both types of errors in mind; that is, all the same good practices apply whether the error is one of omission or commission. However, as a point of emphasis, in assessing the probability of identified potential EOCs, it is particularly important to consider the role that plant indications will play in supporting a crew's ability to detect and recover from EOCs. At least for many EOCs, there may be immediate changes in plant conditions that would alert them to such recovery actions.

6.1.3.2   Good Practice #2: As a Minimum, Search for Conditions That May Make EOCs More Likely

Even if the recommended first good practice above is not performed, the use of risk in any issue assessment should at least ensure that conditions that promote likely EOCs do not exist.
For example, it should be ensured that such conditions have not been introduced by a plant change or modification, or that the plant is not more susceptible to EOCs under the unique set of conditions being examined, (e.g., pressurized thermal shock scenarios). When considering the potential for situations that may make EOCs somewhat likely, the premise of any evaluation should be as follows:

- Operators are performing in a rationale manner (e.g., no sabotage).

- Procedural and training guidance used by the crew will be selected on the basis of the plant status inputs they are receiving.

Using this premise, EOCs are typically the result of problems in the plant information/operating crew interface (e.g., wrong, inadequate information is present, or the information can be easily misinterpreted) or in the procedure-training/operating crew interface (e.g., procedures/training do not cover the actual plant situation very well because they provide ambiguous guidance, no guidance, or even unsafe guidance for the actual situation that may have evolved in a somewhat unexpected way). In either case, significant mismatches can occur between the scenario conditions (i.e., context) and the crew's understanding of those conditions. [See ATHEANA (Ref. 11) for a more detailed discussion on the role of mismatches in facilitating EOCs.]

With a focus on analysts and on reviewers reviewing potential applications of current PRAs, the following is offered as guidance in this area to aid in ensuring EOC-prone conditions do not exist or have not been introduced as part of a plant change. Hence, a review of a plant change should look for situations where one or more of the following characteristics are introduced as a result of the change and, thus, should be corrected if possible:

- To address the potential impacts of mismatches between scenario context and plant information/interface, an analysis/review should at least look for those actions that operators may take that (1) would fail or otherwise make unavailable a PRA function or system, or (2) would reduce the accident mitigating redundancy available, or (3) would exacerbate an accident challenge, because the change has caused such an action to be performed on the basis of only one primary input/indication for which there is no redundant means to verify the true plant status. Such a situation identifies a vulnerable case where EOCs may likely be performed based on only one erroneous (failed, spurious, etc.) input such as an alarm, indicator, or verbal cue of an observed condition.

   In identifying such cases, one should keep in mind that multiple indications may use the same faulty input (e.g., subcooling margin indication and primary system indication may use the same pressure transmitter(s); multiple reactor vessel level indications may rely on the same power supply) and therefore a single fault may actually affect multiple inputs observable to the operator. Depending on how the failure affects the indications (fail high, low, mid-scale, etc.), the failure may not be "obvious" and an EOC-prone situation may exist that may need to be rectified.

- To identify potential problems with the procedure-training interface, an analysis/review should at least look for those actions that operators may take that (1) would fail a PRA function or system, or (2) would reduce the accident mitigating redundancy available, or (3) would exacerbate an accident challenge, because the change has caused the procedure (including entry conditions) and/or training guidance:

   ‣ to become ambiguous/unclear (e.g., vague criteria as to when to abandon the main control room)

   ‣ to introduce a repetitive situation in the response steps where a way to proceed out of the procedure and/or the specific repetitive steps is not evident (e.g., at the end of a series of steps, the procedure calls for a return to a previous step with no clear indication as to how the operators ultimately get out of the repetitive loop of steps)

   ‣ to place the operators in dilemma conditions without some guidance/criteria as to how to "solve" the dilemma (e.g., being vague as to whether or not to shut down a diesel with a cooling malfunction when all other ac power is unavailable)

   ‣ to require the operators to rely on memory, especially for complex or multi-step tasks

   ‣ to require the operators to perform calculations or make other manual adjustments, especially in time-sensitive situations

The above characteristics identify vulnerable cases where EOCs may likely be performed because the procedures and/or training do not adequately cover accident situations that may be faced by the operator or rely on techniques (require memory or adjustments) that may be difficult to perform properly, especially when in a dynamic response situation. In these cases, mismatches between the actual event response that is required and the procedure/training guidance can become magnified making conditions potentially more prone to EOCs.

Note that additional discussions on situations that can facilitate the occurrence of EOCs is provided in ATHEANA (Ref. 11), CESA (Ref. 15), and Julius et al. (Ref. 165).

6.1.4 Possible Impacts of Not Performing Good Practices and Additional Remarks

One concern with failing to adequately address EOCs is that reviews of operational events (e.g., ATHEANA, Ref. 11) have shown that they are often important contributors in serious accidents. Moreover, when they do occur, it is often because the operating crew has become confused because of some unexpected characteristics of the context and there is a strong "basis" for taking the inappropriate action. Thus, not including EOCs will likely cause the overall risk profile to be optimistic since these additional sources of risk will have been neglected. Even though contexts that can lead to EOCs may be relatively rare, when they do occur, there may be a high probability that the EOCs will follow. From a safety standpoint and in terms of identifying potential plant vulnerabilities, it becomes important to at least perform the second good practice above (if the first good practice cannot or is not performed) to at least qualitatively identify those plant conditions that could lead crews to make errors of commission, and address them if necessary.

7. HRA DOCUMENTATION

The ASME Standard (Ref. 5) provides a set of requirements for documenting an HRA in a manner that facilitates PRA applications, upgrades, and peer review. Specific requirements are provided. The following provides good practice for documenting an HRA building on those requirements.

7.1 Documenting the HRA

7.1.1 Objective

The objective is to provide the requirements for documenting an HRA. It is recognized that these requirements may be met at various degrees of detail depending on the application; however, most likely, all of the requirements should be addressed.

7.1.2 Regulatory Guide 1.200 Position

The ASME Standard provides a list of requirements for documenting an HRA under high-level requirement HLR-HR-I, all of which are included in the good practice below. The regulatory guide states that the NRC staff has only one clarification; specifically, this section of the standard is written for impact on CDF and not LERF (seemingly implying that the LERF documentation is covered elsewhere in the standard). Additionally, the reader is referred to Section 1.2.6 of RG 1.200 for further guidance on documentation, in general. That section focuses on the need to ensure traceability and defensibility of the work.

7.1.3.1 Good Practice #1: Document the HRA

The level of detail to be addressed in the documentation depends on the PRA application and the issue being addressed as well as the objectives, scope, and level of detail of the analysis. Whatever documentation is provided, the test for adequate documentation should be whether a knowledgeable reviewer understands the analysis to the point that it can be at least approximately reproduced and the resulting conclusion reached if the same methods, tools, data, key assumptions, and key judgments and justifications are used? Hence, the documentation should include the following, but only to the extent applicable for the application:

- the overall approach and disciplines involved in performing the HRA including to what extent talk-throughs, walkdowns, field observations, and simulations were used

- summary descriptions of the HRA methodologies, processes, and tools used to achieve the following purposes:
  - identify the pre- and post-initiator human actions
  - screen pre-initiators from modeling
  - model the specific HFEs, including decisions about level of detail and the grouping of individual failures into higher order HFEs
  - quantify the HEPs with particular attention to the extent to which plant- and accident sequence-specific information was used, as well as how dependencies were identified and treated

- assumptions and judgments made in the HRA, their bases, and their impact on the results and conclusions (generic or on a HFE-specific basis, as appropriate)

- for at least each of the HFEs important to the risk decision to be made, the PSFs considered, the bases for their inclusion, and how they were used to quantify the HEPs, along with how dependencies among the HFEs and joint probabilities were quantified

- sources of data and related bases or justifications for the following:
  - ▸ screening and conservative values
  - ▸ best estimate values and their uncertainties with related bases

- results of the HRA including a list of the important HFEs and their HEPs

- conclusions of the HRA

7.1.4 Possible Impacts of Not Performing Good Practices and Additional Remarks

Failure to address the topic areas addressed above in the HRA documentation is likely to mean that an outside reviewer or subsequent user of the HRA (peer reviewer, NRC regulator, a new analyst not involved in the original work, etc.) will have insufficient information to be able to independently understand the analysis. In particular, it may be difficult to decide if the good practices have largely been met and whether the HRA results and conclusions are appropriate and defensible. This could be cause for not accepting a proposed plant change or for misuse of the results in other risk-informed decisions (e.g., Maintenance Rule applications).

8. REFERENCES

1.   "Use of Probabilistic Risk Assessment Methods in Nuclear Activities: Final Policy Statement," *Federal Register*, Vol. 60, p. 42622 (60 FR 42622), U.S. Nuclear Regulatory Commission, Washington, DC, August 16, 1995.

2.   *Code of Federal Regulations, Title 10, Parts 1 – 50*, Office of the Federal Register, National Archives and Records Admission, Revised as of January 1, 2001.

3.   "An Approach for Using Probabilistic Risk Assessment in Risk-Informed Decisions on Plant-Specific Changes to the Licensing Basis," Regulatory Guide 1.174, Rev. 1, U.S. Nuclear Regulatory Commission, Washington, DC, November 2002.

4.   "An Approach for Determining the Technical Adequacy of Probabilistic Risk Assessment Results for Risk-Informed Activities," Draft Regulatory Guide 1.200, U.S. Nuclear Regulatory Commission, Washington, DC, February 2004.

5.   "Standard for Probabilistic Risk Assessment for Nuclear Power Plant Applications," ASME RA-S-2002 (which includes the Addenda to the Standard, RA-Sa-2003), American Society of Mechanical Engineers, April 5, 2002.

6.   "Probabilistic Risk Assessment Peer Review Process Guidance," NEI 00-02, Revision A3, Nuclear Energy Institute, March 2000

7.   "Use of Probabilistic Risk Assessment in Plant-Specific, Risk-Informed Decisionmaking: General Guidance," NUREG-0800, Chapter 19, Rev. 1, U.S. Nuclear Regulatory Commission, Washington, DC, November 2002.

8.   "Guidance for the Review of Changes to Human Actions," NUREG-1764, U.S. Nuclear Regulatory Commission, Washington, DC, December 2002.

9.   Swain, A.D., and H.E. Guttmann, "Handbook of Human Reliability Analysis with Emphasis on Nuclear Power Plant Applications," NUREG/CR-1278, SAND80-0200, Sandia National Laboratories, August 1983.

10.   Embrey, D.E., et al., "SLIM-MAUD: An Approach to Assessing Human Error Probabilities Using Structured Expert Judgment," NUREG/CR-3518, Vols. I and II, Brookhaven National Laboratory, Upton, NY, 1984.

11.   "Technical Basis and Implementation Guidelines for A Technique for Human Event Analysis (ATHEANA)," NUREG-1624, Rev. 1, U.S. Nuclear Regulatory Commission, Washington, DC, May 2000.

12.   "Severe Accident Risks: An Assessment for Five U.S. Nuclear Power Plants," NUREG-1150, U.S. Nuclear Regulatory Commission, Washington, DC, December 1990.

13. Bieder, C., P. Le-Bot, E. Desmares, J-L Bonnet, F. Cara, "MERMOS: EDF's New Advanced HRA Method," in *Probabilistic Safety Assessment and Management (PSAM 4),* A. Mosleh and R.A. Bari (editors), Springer-Verlag, New York, 1998.

14. Hollnagel, E., *Cognitive Reliability and Error Analysis Method (CREAM).* Elsevier Science, New York, 1998.

15. Reer, B., V.N. Dang, and S. Hirschberg, "The CESA Method and Its Application in a Plant-Specific Pilot Study on Errors of Commission," *Reliability Engineering & System Safety, 83:* 187–205, 2004.

16. Julius, J., E. Jorgenson, G.W. Parry, and A.M. Mosleh, "A Procedure for the Analysis of Error of Commission in a Probabilistic Safety Assessment of a Nuclear Power Plant at Full Power," *Reliability Engineering and System Safety, 50:* 189–201, 1995.

17. Macwan, A., and A. Mosleh, "A Methodology for Modeling Operator Errors of Commission in Probabilistic Risk Assessment," *Reliability Engineering & System Safety, 45:* 139–157, 1994.

18. Vuorio, U.M., and J.K. Vaurio, "Advanced Human Reliability Analysis Methodology and Applications," In: *Proceedings of PSA '87: Probabilistic Safety Assessment International Meeting, Zurich, Switzerland, August 3 – September 4, 1987,* pp.104–109, American Nuclear Society, La Grange Park, IL, 1987.

19. Wakefield, D.J., "Application of the Human Cognitive Reliability Model and Confusion Matrix Approach in a Probabilistic Risk Assessment," *Reliability Engineering and System Safety, 22:* 295–312, 1988.

20. Davoudian, K., J.S. Wu, and G. Apostolakis, "Incorporating Organizational Factors into Risk Assessment through the Analysis of Work Processes," *Reliability Engineering and System Safety,* 45: 85–105, 1994.

21. Davoudian, K., J.S. Wu, and G. Apostolakis, "The Work Process Analysis Model (WPAM)," *Reliability Engineering and System Safety,* 45: 107–125, 1994.

22. Singh, A.J., G.W. Parry, and A.N. Beare, "An Approach to the Analysis of Operating Crew Response for Use in PRAs," In: *Proceedings of PSA '93: Probabilistic Safety Assessment International Meeting, Clearwater Beach, FL, January 27–29, 1993,* pp.294–300, American Nuclear Society: La Grange Park, IL, 1993.

23. Reason, J., *Human Error,* Cambridge, England: Cambridge University Press, 1990 and *Managing the Risks of Organizational Accidents,* Aldershot, UK: Ashgate, 1997.

24. Woods, D.D., et. al., *Behind Human Error: Cognitive Systems, Computers, and Hindsight,* Crew System Ergonomics Information Analysis Center (CSERIAC), The Ohio State University, Wright-Patterson Air Force Base, Columbus, OH, December 1994.

25. Endsley, M.R., "Towards a Theory of Situation Awareness in Dynamic Systems," *Human Factors, 37:* 65–84, 1995.

26. Sträter, O., and H. Bubb, "Design of Systems in Settings with Remote Access to Cognitive Performance," Hollnagel, E. (Editor), *Handbook of Cognitive Task Design*, Lawrence Erlbaum, Mahwah, NJ, 2003.

27. Baranowsky, P.D., G. Rasmuson, A. Johanson, A. Kreuser, P. Pyy, and W. Werner, "General Insights from the International Common Cause Failure Data Exchange (ICDE) Project," *Probabilistic Safety Assessment and Management (PSAM 7)*, C. Spitzer, U. Schmocker, and V.N. Dang (Editors), Springer-Verlag, London Limited, 2004.

28. Tirira, J., and W. Werner, "Lessons Learned from Data Collected in the ICDE Project," *Probabilistic Safety Assessment and Management (PSAM 7)*, C. Spitzer, U. Schmocker, and V.N. Dang (Editors), Springer-Verlag, London Limited, 2004.

29. Johanson, G., P. Hellstrom, T. Makamo, J-P. Bento, M. Knochenhauer, and K. Porn, Dependency Defense and Dependency Analysis Guidance, Swedish Nuclear Power Inspectorate (SKI) Research Report (SKI 0404), October 2003.

30. Swain, A.D., "Accident Sequence Evaluation Program Human Reliability Analysis Procedure," NUREG/CR-4772, SAND86-1996, Sandia National Laboratories, February 1987.

31. "Probabilistic Risk Assessment Reference Document," NUREG-1050, U.S. Nuclear Regulatory Commission, Washington, DC, 1984.

APPENDIX A.

SUMMARY OF THE HRA GOOD PRACTICES AUDIT

APPENDIX A.
SUMMARY OF THE HRA GOOD PRACTICES AUDIT

An audit of HRA good practices was conducted against the supporting requirements (SRs) in Section 4.5.5, "Human Reliability Analysis," of ASME RA-Sa-2003 and Table HR, "Human Reliability Analysis (HRA) Modeling Related Grades: Element HR," in Appendix B of NEI 00-02. Other SRs and technical elements (TEs) relevant to HRA practice may be contained in other sections of these references; however, these were neither identified nor considered in this review.

The audit indicated that draft HRA good practices existed for all SRs in Section 4.5.5 of ASME RA-Sa-2003, and all TEs in Table HR of NEI 00-02, with the exception of HR-29 pertaining to the need for an independent review of the HRA results. Note that the empty cells in the RG 1.200 column indicate that the NRC had no objection to the SRs in Section 4.5.5 of ASME RA-Sa-2003, or the TEs in Table HR of NEI 00-02.

Table A-1						
Summary of HRA Good Practices Audit Against ASME RA-Sa-2003 and NEI 00-02						
Location	GP	Description	ASME RA-Sa-2003	NEI 00-02	RG 1.200	Notes
HRA team formation and techniques for a realistic analysis						
Section 3.1.3.1	1	Perform a Multi-Disciplinary, Integrated Analysis	not addressed	not addressed		
Section 3.1.3.2	2	Perform Field Observations and Discussions	HR-E3	HR-10 HR-14 HR-20		
Identifying potential pre-initiator human failure events (HFEs)						
Section 4.1.3.1	1	Review Pre-Initiator Procedures, Actions, and Equipment	HR-A1 HR-A2 HR-D3	HR-4 HR-5	clarification to NEI 00-02	Self-assessment needs to confirm and document that the factors listed in ASME HR-D3 were considered in the pre-initiator HEPs
Section 4.1.3.2	2	Identify All Pre-Initiator Human Actions (Do Not Ignore Pre-Initiators)	HR-C2 HR-C3	HR-5 HR-7 HR-27		
Section 4.1.3.3	3	Examine Other Operational Modes and Routine Actions	not addressed	not addressed		
Section 4.1.3.4	4	Identify Actions Affecting Redundant and Multiple Diverse Equipment	HR-A3 HR-B2	HR-5 HR-6 HR-7 HR-26		

Table A-1
Summary of HRA Good Practices Audit Against ASME RA-Sa-2003 and NEI 00-02

Location	GP	Description	ASME RA-Sa-2003	NEI 00-02	RG 1.200	Notes
	**Screening those pre-initiator activities for which HFEs do not need to be modeled**
Section 4.2.3.1	1	Screen Pre-Initiators with Acceptable Restoration Mechanisms or Aids	HR-B1	HR-5 HR-6		
Section 4.2.3.2	2	Do Not Screen Actions Affecting Redundant and Multiple Diverse Equipment	HR-A3 HR-B2	HR-5		
Section 4.2.3.3	3	Reevaluate the Screening Process for Special Applications	Section 3	not addressed		Section 3 of the ASME standard provides detailed discussion about the risk assessment application process
	**Modeling specific HFEs corresponding to the unscreened pre-initiator human actions**
Section 4.3.3.1	1	Include HFEs for Unscreened Human Actions in the PRA Model.	HR-C1 HR-C2 HR-C3	HR-5 HR-7 HR-27		
	**Quantifying the corresponding HEPs for the specific pre-initiator HFEs**
Section 4.4.3.1	1	Use Screening Values During the Initial Quantification of the HFEs	HR-B1 HR-D1 HR-D2	HR-5 HR-6 HR-13		
Section 4.4.3.2	2	Detailed assessments of at least the significant HFE contributors should be performed	HR-B1 HR-D1 HR-D2	HR-5 HR-6		
Section 4.4.3.3	3	Revisit the use of screening vs. detailed-assessed HEPs for specific PRA applications	Section 3	not addressed		Section 3 of the ASME Standard provides detailed discussion about the risk assessment application process
Section 4.4.3.4	4	Account for Plant- and Activity-Specific Performance-Shaping Factors (PSFs) in the Detailed Assessments	HR-G3 HR-G4 HR-G5	HR-6		
Section 4.4.3.5	5	Apply Plant-Specific Recovery Factors	HR-D4	not addressed	clarification to NEI 00-02	NEI 00-02 does not address use of expert judgment

Table A-1
Summary of HRA Good Practices Audit Against ASME RA-Sa-2003 and NEI 00-02

Location	GP	Description	ASME RA-Sa-2003	NEI 00-02	RG 1.200	Notes
Section 4.4.3.6	6	Account for Dependencies Among the HEPs in an Accident Sequence	HR-D5	HR-26 HR-27		
Section 4.4.3.7	7	Assess the Uncertainty in HEPs	HR-D6	not addressed		
Section 4.4.3.8	8	Evaluate the Reasonableness of the HEPs Obtained Using Detailed Assessments	HR-D7	not addressed		
Identifying potential post-initiator human failures						
Section 5.1.3.1	1	Review Post-Initiator Related Procedures and Training Materials (and plant simulator training)	HR-E1(a) HR-E3 HR-E4	HR-9 HR-10 HR-14 HR-16 HR-20		
Section 5.1.3.2	2	Review Functions and Associated Systems and Equipment to be Modeled in the PRA	HR-E1(b)	HR-9 HR-10 HR-16		
Section 5.1.3.3	3	Look for Certain Expected Types of Actions (general types of post-initiator actions to be considered)	HR-E2	HR-8 HR-9 HR-10 HR-21 HR-22 HR-23 HR-25		
Modeling specific HFEs corresponding to the post-initiator human actions						
Section 5.2.3.1	1	Include HFEs for Needed Human Actions in the PRA Model	HR-F1	HR-16		
Section 5.2.3.2	2	Define the HFEs Such that they are Plant- and Accident Sequence-Specific	HR-F2	HR-11 HR-16 HR-17 HR-19 HR-20		
5.2.3.3	3	Perform Talk-Throughs, Walkdowns, Field Observations, and Simulator Exercises (as necessary) to Support the Modeling of Specific HFEs	HR-E3 HR-E4 HR-G5	HR-10 HR-14 HR-16 HR-18 HR-20		

			Table A-1			
		Summary of HRA Good Practices Audit Against ASME RA-Sa-2003 and NEI 00-02				
Location	**GP**	**Description**	**ASME RA-Sa-2003**	**NEI 00-02**	**RG 1.200**	**Notes**
Quantifying the corresponding HEPs for the specific post-initiator HFEs						
Section 5.3.3.1	1	Address Both Diagnosis and Response Execution Failures	HR-G2	HR-2 HR-11	Qualification to NEI 00-02	Self-assessment needs to document if diagnosis and execution errors are included
Section 5.3.3.2	2	Use Screening Values During the Initial Quantification of the Post-Initiator HFEs	HR-G1	HR-13 HR-15 HR-17 HR-18		
Section 5.3.3.3	3	Perform Detailed Assessments of Significant Post-Initiator HFEs.	HR-G1	HR-15 HR-17 HR-18		
Section 5.3.3.4	4	Revisit the Use of Post-Initiator Screening Values vs. Detailed Assessments for Special PRA	Section 3	not addressed		Section 3 of the ASME Standard provides detailed discussion about the risk assessment application process
Section 5.3.3.5	5	Account for Plant- and Activity-Specific PSFs in the Detailed Assessments of Post-Initiator HEPs	HR-G3 HR-G4 HR-G5	HR-16 HR-17 HR-18 HR-19 HR-20		
Section 5.3.3.6	6	Account for Dependencies Among Post-Initiator HFEs	HR-G7 HR-G8	HR-26 HR-27	Clarification to ASME HR-G7	RG 1.200 added a requirement to justify multiple recovery actions
Section 5.3.3.7	7	Assess the Uncertainty in HEPs	HR-G9	not addressed		
Section 5.3.3.8	8	Evaluate the Reasonableness of the HEPs Obtained Using Detailed Assessments	HR-G6	HR-12		
Adding recovery actions to the PRA						
Section 5.4.3.1	1	Define Appropriate Recovery Actions	HR-H1 HR-H2	HR-21 HR-22 HR-23 HR-24		
Section 5.4.3.2	2	Account for Dependencies (all earlier guidance applies for recovery actions;	HR-H3	HR-26	Clarification to ASME HR-H3	RG 1.200 added a requirement to justify multiple recovery actions

Table A-1 Summary of HRA Good Practices Audit Against ASME RA-Sa-2003 and NEI 00-02						
Location	GP	Description	ASME RA-Sa-2003	NEI 00-02	RG 1.200	Notes
Section 5.4.3.3	3	Quantify the Probability of Failing to Perform the Recovery(ies).	Quantifca-tion not explicitly addressed	Quantifca-tion not explicitly addressed		
Including and modeling errors of commission (EOCs)						
Section 6.1.3.1	1	Address EOCs in Future HRAs/PRAs (Recommendation).	not addressed	not addressed		
Section 6.1.3.2	2	As a Minimum, Search for Conditions that May Make EOCs More Likely.	not addressed	not addressed		
Documenting the HRA						
Section 7.1.3.1	1	HRA documentation	HR-I1	HR-1 HR-28 HR-30	Clarification to ASME HR-I1	RG 1.200 maintains that the ASME SRs pertaining to HRA "are written for CDF and not LERF"

APPENDIX B.

GUIDANCE ON CONSIDERATION
OF PERFORMANCE-SHAPING FACTORS
FOR POST-INITIATOR HFEs

APPENDIX B.
GUIDANCE ON CONSIDERATION
OF PERFORMANCE-SHAPING FACTORS
FOR POST-INITIATOR HFEs

This appendix provides additional detail concerning the performance-shaping factors (PSFs) presented in Section 5.3.3.5, including some key characteristics to consider when assessing the influence of these PSFs on the failure probability for a human failure event (HFE). Included are important interactions among the factors that should also be examined when assessing the holistic impact of the PSFs on operator performance. These factors need to be assessed on a plant- and accident sequence-specific basis, considering the relevant context and the action to be performed.

It is important to reiterate that this Appendix is written for the specific purpose of addressing post-initiator HFEs in a risk assessment for commercial nuclear power plant (NPP) operations occurring nominally at full power, and for internal initiating events. However, much of it is considered useful to other modes of operation and for other industry applications such as safety assessments of chemical plants, space mission risk assessments, and others. Similarly, much of it is considered applicable for external initiating events, but it should be used with the additional context of such events in mind (e.g., shaking during a seismic event). Additionally, portions of this appendix may be of benefit in examining human actions related to nuclear materials and safeguard types of applications.

Specific HRA methods and tools used by the industry may define and "measure" these PSFs somewhat differently than described here. That is, they may use a different explicit set of PSFs that "roll up" many of the factors listed below into the definitions of their specific factors (e.g., stress, workload). Nonetheless, these summaries are provided as one means with which to assess that the specific HRA method/tool has been used such that the characteristics described here have indeed been accounted for in the evaluation of post-initiator human error probabilities (HEPs).

While quantitative guidance is not provided (specific quantification depends on the method/tool that is used), the following should be useful in determining qualitatively whether a PSF is a weak/strong positive, neutral (or not applicable), or negative influence, regardless of the method/tool. The method/tool being used should provide definitions and guidance for assessing PSFs qualitatively (e.g., "good," "adequate," "poor"), along with a way to interpret the results into a quantified HEP, that can be used in conjunction with this information.

The PSFs are addressed below.

B.1 Applicability and Suitability of Training/Experience

For both in-control room and local actions, this is an important factor in assessing operator performance. For the most part, in nuclear plants, operators can be considered "trained at some minimum level" to perform their desired tasks.

However, from an HRA perspective, the degree of familiarity with the type of sequences modeled in the PRA and the actions to be performed, can provide a negative or positive influence that should be used to assess the likelihood of operator success. In cases where the type of PRA sequence being examined or the actions to be taken are not periodically addressed in training (such as covered in classroom sessions or simulated every 1 to 2 years or even more often) or the actions are not performed as part of their normal experience or on-the-job duties, this factor should be treated as a negative influence. The converse would result in a positive influence on overall operator performance.

One should also attempt to identify systematic training biases that may affect operator performance either positively or negatively. For example, training guidance in a pressurized-water reactor (PWR) may provide a reluctance to use "feed-and-bleed" in a situation where steam generator feed is expected to be recovered. Other biases may suggest operators are allowed to take certain actions before the procedural steps calling for those actions are reached, if the operators are sure the actions are needed. Such training "biases" could cause hesitation and result in higher HEPs for the desired actions, as in the first case above, or as in the case of not waiting to take obvious actions, they may be a positive influence.

It is incumbent on the analyst to ensure that training and/or experience is relevant to the PRA sequence situation and desired actions. The more it can be argued that the training is current, "is like the real event," is varied enough to represent differences in the way the event can evolve, and proficiency is demonstrated on a periodic basis, the more positive this factor. If there is little or no training/experience or there are potentially negative training biases for the PRA sequence being examined, this factor should be considered to have a negative influence.

B.2 Suitability of Relevant Procedures and Administrative Controls

For both in-control room and local actions, this is an important factor in assessing operator performance. Similar to training, for the most part, procedures exist to cover many types of sequences and operator actions.

However, from an HRA perspective, the degree to which the procedures clearly and unambiguously address the types of sequences modeled in the PRA and the actions to be performed, dictates whether they are a negative or positive influence on operator performance. Where procedures have characteristics like those listed below related to the desired actions for the sequences of interest, this factor should be considered a negative influence:

- ambiguous, unclear, or non-detailed steps for the desired actions in the context of the sequence of interest

- situations where the operators are likely to have trouble identifying a way to proceed forward through the procedure

- requirements to rely on considerable memory

- situations in which operators must perform calculations or make other manual adjustments (especially time-sensitive situations)

- situations for which there is no procedure, or the procedure is likely to not be available, especially when taking local actions "in the heat of the scenario" and it cannot be argued that the desired task is simple and a "skill of the craft" or that it is an automatic/memorized activity that is trained on and for which there is routine experience

- "double negatives" are present in the procedures (they should be evaluated to determine whether there are some circumstances which could make them particularly confusing)

Otherwise, this PSF should be considered as adequate or even a positive influence.

Talk-throughs with operations and training staff (in the context of the scenario being examined) can be helpful in uncovering "difficulties" or "ease" in using the relevant procedures, considering the associated training that the operators receive and the way the operators interpret the use of the procedures.

B.3   Availability and Clarity of Instrumentation (Cues to Take Actions and Confirm Expected Plant Response)

For both in-control room and local actions, this is an important factor since operators, other than for immediate and memorized response actions, take actions based on diagnostic indications and look for expected plant responses to dictate follow-on actions. For in-control room situations, typical nuclear plant control rooms have sufficient redundancy and diversity for most important plant parameters. For this reason, most HRA methods inherently assume that adequate instrumentation typically exists. Nonetheless, this should be verified looking for the following characteristics that could make this a negative PSF, particularly in situations where there is little redundancy in the instrumentation associated with the act(s) of interest:

- the key instrumentation associated with an action is adversely affected by the initiating event or subsequent equipment failure (e.g., loss of DC power causing loss of some indications, spurious or failed as a result of a hot short from a fire)

- the key instrumentation is not readily available and may not be typically scanned such as on an obscure back panel

- the instrumentation could be misunderstood or may be ambiguous because it is not a direct indication of the equipment status (e.g., PORV position is really the position of the solenoid valve and not the PORV itself)

- the instrumentation is operating under conditions for which it is not appropriate (e.g., calibrated for normal power conditions as opposed to shutdown conditions)

- there are so many simultaneous changing indications and alarms or the indication is so subtle, particularly when the time to act is short, it may be difficult to "see and pick out" the important cue in time (e.g., a changing open-close light for a valve without a concurrent alarm or other indication, finding one alarm light among hundreds).

The above also applies to local actions outside the control room, recognizing that in some situations, less instrumentation may exist (e.g., only one division of instrumentation and limited device actuators on the remote shutdown panel). However, on the positive side, local action indications often can include actual/physical observation of the equipment (e.g., pump is running, valve stem shows it is closed) that compensates for any lack of other indicators or alarms.

It is incumbent on the analyst to ensure that adequate instrumentation is available and clear so that the operators will know the status of the plant and when certain actions need to be taken. If this is demonstrated, then this PSF would be positive. Task analysis will often facilitate determining whether the instrumentation is adequate.

B.4   Time Available and Time Required to Complete the Act, Including the Impact of Concurrent and Competing Activities

This can be an important influence for both in-control room and local actions since clearly, if there is not enough or barely enough time to act, the estimated HEP is expected to be quite high. Conversely, if the time available far exceeds the time required and there are not multiple competing tasks, the estimated HEP is not expected to be strongly influenced by this factor.

It is important that the time available and the time needed to perform the action be considered *in concert with* many of the other PSFs and the demands of the sequence. This is because the thermal-hydraulic inputs (e.g., time to steam generator dryout, time to start uncovering the core), while important, are not the only influences. (Note, it is best if the thermal-hydraulic influences are derived from plant-specific or similar analyses rather than simple judgments).

The time to perform the act, in particular, is a function of the following factors:
• number of available staff
• clarity and repetitiveness of the cues that the action needs to be performed
• the human-system interface (HSI, discussed later)
• complexity involved (discussed later)
• need to get special tools or clothing (discussed later)
• consideration of diversions and other concurrent requirements (discussed later)
• where in the procedures the steps for the action of interest are called out
• crew characteristics such as whether the crews are generally aggressive or slow and methodical in getting through the procedural steps (discussed later)
• other potential "time sinks"

Clearly there is judgment involved, but as described here, it is not as simple as watching an operator perform an action in ideal conditions with a stop watch to determine the time required to perform the act. Only when the sequence context is considered holistically with the interfacing PSFs that have been mentioned here, can more meaningful "times" be estimated. Hence, especially for complex actions and/or situations, walkdowns and simulations can be helpful in ensuring overly optimistic "times" have not been estimated. Whatever HRA method/tool is used, determination of these times should include the considerations provided here.

B.5 Complexity of the Required Diagnosis and Response, the Need for Special Sequencing, and the Familiarity of the Situation

This factor attempts to measure the overall complexity involved for the situation at hand and for the action itself (e.g., many steps have to be performed by the same operator in rapid succession vs. one simple skill-of-the-craft action). Many of the other PSFs bear on the overall complexity, such as the need to decipher numerous indications and alarms, the presence of many and complicated steps in a procedure, poor HSI, etc. Nonetheless, this factor should also capture "measures," such as the ambiguity associated with assessing the situation or in executing the task, the degree of mental effort or knowledge involved, whether it is a multi-variable or single-variable associated task, whether special sequencing or coordination is required in order for the action to be successful (especially if it involves multiple persons in different locations), whether the activity may require very sensitive and careful manipulations by the operator, etc. The more these "measures" describe an overall complex situation, this PSF should be found to be a negative influence. To the extent these "measures" suggest a simple, straightforward, unambiguous process (or one that the crew or individual is very familiar with and skilled at performing), this factor should be found to be nominal or even ideal (i.e., positive influence).

B.6 Workload, Time Pressure, and Stress

Although these factors are often associated with complexity and can certainly contribute to perceived complexity, the emphasis here is on the amount of work a crew or individual has to accomplish in the time available (e.g., task load), along with their overall sense of being pressured and/or threatened in some way with respect to what they are trying to accomplish [e.g., see Swain and Guttmann (Ref. 9) for a more detailed definition and discussion of stress and workload]. To the extent crews or individuals expect to be under high workload, time pressure, and stress, it is generally thought to have a negative impact on performance (particularly if the task being performed is considered complex). However, the impact of these factors should be carefully considered in the context of the scenario and the other PSFs thought to be relevant. For example, if the scenario is familiar, the procedures and training for the scenario are very good, and the rate at which the crews normally implement their procedures will allow them to achieve their goal on time, then analysts might decide that even relatively high expected levels of workload and stress will not have a significant impact on performance. Although these factors may be difficult to measure, analysts should demonstrate a careful evaluation of their potential influence in the scenario being examined, before deciding on the strength of their effect.

B.7 Team/Crew Dynamics and Crew Characteristics [Degree of Independence Among Individuals, Operator Attitudes/Biases/Rules, Use of Status Checks, Approach for Implementing Procedures (e.g., Aggressive Crew vs. Slow/Methodical Crew)]

This is a "catch-all" type of factor which can be important, especially to in-control room actions where the early responses to an event occur and the overall strategy for dealing with the event develops. In particular, the way the procedures are written and what is (or is not) emphasized in training (which may be related to an organizational or administrative influence), can cause systematic and nearly homogeneous biases and attitudes in most or all the crews that can affect overall crew performance. A review of this factor should include the following characteristics (among others):

- Are independent actions encouraged or discouraged among crew members (allowing independent actions may shorten response time but could cause inappropriate actions going unnoticed until much later in the scenario)?

- Are there common biases or "informal rules"? For example, is there a reluctance to do certain acts, is there an overall philosophy to protect equipment or run it to destruction if necessary, or are there informal rules regarding the way procedural steps are interpreted.

- Are periodic status checks performed (or not) by most crews so that everyone has a chance to "get on the same page" and allow for checking on what has been performed to ensure that the desired activities have taken place? In general, are there good communication strategies used to help ensure that everyone stays informed?

- Is the overall approach of most crews to aggressively respond to the event, including taking allowed shortcuts through the procedural steps (which will shorten response times), or are typical responses slow and methodical ("we trust the procedures" type of attitude), thereby tending to slow down response times but making it less likely to make mistakes.

In general, deciding whether the crew characteristics have a positive or negative effect will be contingent on the scenario being examined. For example, a particular bias may be very positive for some scenarios, but not for others.

Observing simulations and using talk-throughs and walkdowns can provide valuable insights into the overall crew response dynamics, attitudes, and the typical times it takes them to get through various procedure steps and deal with unexpected failures or distractions (also Section 3.1.3.2). This knowledge can be a key input into the HEP evaluation including determining the typical time to respond (see that factor above).

B.8   Available Staffing/Resources

For in-control room actions, this is generally not an important consideration (i.e., not a particularly positive or negative factor) since plants are supposed to maintain an assigned minimum crew with the appropriate qualified staff available in or very near the control room.

However, for ex-control room local actions, this can be an important consideration particularly dependent on (1) the number and locations of the necessary actions, (2) the overall complexity of the actions that must be taken, and (3) the time available to take the actions and the time required to perform the actions (see above for more on these related factors). For instance, where the actions are few, complexity is low, and available time is high, one or two personnel available to perform the local actions may be more than enough and thus the available staffing can be considered to be adequate or even a positive factor. On the other hand, where the number of actions and their complexity is high, and with little time, perhaps three or more staff may be necessary. Additionally, the time of the day the initiating event occurs may be a factor since typically, night and "back" shifts have fewer people available than the day shift. If there are significant differences in the staffing levels depending on the time of day, it is advisable to either treat the staffing level in an HEP evaluation as the minimum available depending on the shift, or probabilistically account for these aleatory differences more explicitly in the PRA model. *It is incumbent on the analyst to demonstrate that the available staffing is sufficient to perform the desired actions and/or assess the HEP(s) accordingly.*

B.9   Ergonomic Quality of the Human-System Interface (HSI)

This is generally not an important factor relative to main control room actions since, given the many control room design reviews and improvements and the daily familiarity of the control room boards and layout, problematic HSIs have been taken care of or are easily worked around by the operating crew. Of course, any known very poor HSI should be considered as a negative influence for an applicable action even in the control room. For example, if common workarounds are known to exist that may negatively influence a desired act, this should be accounted for in the HEP evaluation. Furthermore, it is possible that some unique situations may render certain HSIs less appropriate and for such sequences, the relevant interfaces should be examined.

However, since local actions may involve more varied (and not particularly "human-factored") layouts and require operators to take actions in much less familiar surroundings and situations, any problematic HSIs can be an important negative factor on operator success. For instance, if to reach a valve to open it manually requires the operator to climb over pipes and turn the valve with a tool while in a laid out position, or in-field labeling of equipment is generally in poor condition and could lengthen the time to find the equipment, etc., such "less ideal" HSIs could mean this is a negative performance-shaping factor. Otherwise, if a review reveals no such problematic interfaces for the act(s) of interest, this influence can be considered adequate, or even positive if the interface helps ensure the appropriate response in some way.

Walkdowns and field or simulator observations can be useful tools in discovering problems (if any exist) in the HSI for the actions of interest. Sometimes, discussions with the operators will reveal their own concerns about issues in this area.

B.10 Environment in Which the Action Needs To Be Performed

Except for relatively rare situations, this factor is not particularly relevant to in-control room actions, given the habitability control of such rooms and the rare challenges to that habitability (e.g., control room fire, loss of control room ventilation, less lighting as a result of station blackout). However, for local actions, this could be an important influence on the operator performance. Radiation, lighting, temperature, humidity, noise level, smoke, toxic gas, even weather for outside activities (e.g., having to go on a potential snow-covered roof to reach the atmospheric dump valve isolation valve), etc., can be varied and far less than ideal. Hence any HEP assessment should ensure that the influence of the environment where the act(s) needs to take place is accounted for as a performance-shaping factor. This factor may be non-problematic (adequate) or a negative influence (even to the point of not being able to perform the act).

B.11 Accessability and Operability of the Equipment To Be Manipulated

As with the environment factor, this factor is not particularly relevant most of the time to in-control room actions except for special circumstances such as loss of operability of indications or controls as a result of the initiator or equipment failures (e.g., loss of DC). However, for local actions, accessability and the operability of the equipment to be manipulated may not always be ensured, and needs to be assessed in the context with such influences as the environment, the need to use special equipment (discussed later), and HSI. Hence any HEP assessment should ensure that this factor, for where the act(s) needs to take place, is accounted for as a performance-shaping factor. This factor may be non-problematic (adequate) or a negative influence (even to the point of not being able to perform the act).

B.12 Need for Special Tools (Keys, Ladders, Hoses, Clothing Such as To Enter a Radiation Area)

As for the environment and accessability factors, this factor is not particularly relevant to in-control room actions with the common exception of needing keys to manipulate certain control board switches or similar controls (e.g., key for explosive valves for standby liquid control injection in a BWR). However, for local actions, such needs may be more commonplace and necessary in order to successfully perform the desired act. If such equipment is needed, it should be ensured that the equipment is readily available, its location is readily known, and it is either easy to use or periodic training is provided, in order for this factor to be considered to be adequate. Otherwise, this factor should be considered to have a negative influence on the operator performance, perhaps even to the point of making the failure of the desired action very high.

B.13 Communications (Strategy and Coordination) and Whether One Can Be Easily Heard

For actions in the control room, this factor is not particularly relevant, although there should be verification that the strategy for communicating in the control room is one that tends to ensure that directives are not easily misunderstood (e.g., it is required that the board operator repeat the action to be performed and then wait for confirmation before taking the act). Do crew members avoid the use of double negatives? Generally, it is expected that communication will not be problematic; however, any potential problems in this area (such as having to talk with special air packs and masks on in the control room in a minor fire) should be accounted for if they exist.

For local actions, this factor may be much more important because of the possible less than ideal environment or situation. It should be assured that the initiating event (e.g., loss of power, fire, seismic) or subsequent equipment faults are not likely to negatively affect the ability for operators to communicate as necessary to perform the desired act(s). For instance, having to set up the equipment and talk over significant background noise and possibly having to repeat oneself many times should be a consideration — even if only as a possible "time sink" for the time to perform the act. Additionally, there should be training on the use of the communication equipment, its location should be readily known, and its operability periodically demonstrated and shown to be in good working condition. Depending on the status of these characteristics, this factor may be non-problematic (adequate) or a negative influence (even to the point of not being able to perform the act).

B.14 Special Fitness Needs

While typically not an issue for in-control room actions, this could be an important factor for a few local actions depending on the specific activity involved. Having to climb up or over equipment to reach a device, needing to move and connect hoses, using an especially heavy or awkward tool, are examples of where this factor could have some influence on the operator performance. In particular, the response time for an action may be increased for successful performance of the act. Physically demanding (or not) activities should be considered in the evaluation of any HEP where it is appropriate to do so. Talk-throughs or field observations of the activities involved can help determine whether such issues are relevant to a particular HFE.

B.15 Consideration of "Realistic" Accident Sequence Diversions and Deviations (e.g., Extraneous Alarms, Outside Discussions, or the Sequence Evolution Is Not Exactly Like That Trained On)

Particularly for in-control room actions where the early responses to an event occur and the overall strategy for dealing with the event develops, this can be an important factor to be considered. Through simulations, training, and the way the procedures are written, operators "build up" some sense of expectations as to how various types of sequences are likely to proceed; even to the extent of recognizing alarm and indication patterns and what actions will likely be appropriate. To the extent the actual sequence may not be "just like in the simulator," such as involving other unimportant or spurious alarms, the need for outside discussions with other staff or even offsite entities such as a fire department, differences in the timing of the failed events, and behavior of critical parameters, etc., all can add to the potential diversions and distractions that may delay response timing or in the extreme, even confuse the operators as to the appropriate actions to take.

Hence, the "signature" of the PRA accident sequence and the potential actions of interest should be examined against the expectations of the operators to determine if there is a considerable potential for such distractions and deviations. Observing simulations and talking with the operators can help in discovering such possibilities. This could impact the HEP mean value estimate as well as the uncertainty in the HEP, which may be important to assessing the potential risk or in establishing the limits for doing sensitivity studies with the HEP.

APPENDIX C.

**SUMMARY OF PUBLIC COMMENTS
ON THE AUGUST 2004 DRAFT OF NUREG-1792**

APPENDIX C.
SUMMARY OF PUBLIC COMMENTS
ON THE AUGUST 2004 DRAFT OF NUREG-1792

This appendix summarizes the public comments received on the August 2004 draft of this report. Each specific comment is not detailed here; however, this summary reflects the nature of all comments that the NRC received, with the agency's related responses as noted.

Comment #1
The NRC received comments addressing the need to clarify the target audience of the report, the need for the report, concerns about the report becoming de facto requirements, related concerns about controlling the implementation of the report, and how these practices relate to the capability categories in ASME Standard RA-S-2002.

Response #1
As stated in this report, Regulatory Guide 1.200 and the related ASME Standard RA-S-2002 (with the RA-Sa-2003 addenda) and NEI 00-02, provide the standards to be met, but do not go far enough in describing reasonably acceptable practices for meeting those standards. As the use of probabilistic risk assessment (PRA) and risk-informed analyses becomes even more prevalent in the industry, human reliability analyses (HRAs) remain one of the areas for which there is less of a consensus as to how to perform the analyses and, thus, continue to be among the more uncertain portions of risk assessments. As a result, particularly for internal use by the NRC staff, but also having implications for those performing and submitting HRAs for NRC consideration, this report provides additional guidance for reasonable steps to take in order to meet the standards.

This report does not constitute a standard and, hence, it is not intended to provide de facto requirements. As previously stated in the report, and with additional clarification added as a result of these comments, this report is intended to serve as a reference guide for widely accepted good practices in HRA that are currently available to cover the breadth of possible future HRA applications. The report clearly states that all good practices need not always be met; the nature of the issue being addressed will determine which practices are applicable, and the degree to which those practices should be applied. Any technical analysis should justify the appropriateness of the scope and level of detail of the analysis. That justification should be part of the analysis documentation and it will largely serve to limit the degree that these practices should be applied. In that way, both analysts and reviewers should understand that certain practices are not applicable or are sufficiently met even if applied to only a limited degree.

As a result, the good practices herein are not tied to specific capability categories in the ASME Standard. Depending on the issue being addressed, the analysts should decide what capability category requirements apply (this establishes which requirements in the standard apply), and then the corresponding good practices should be applied as appropriate, based on the analysis justification about scope and level of detail.

This report is intended to be used as described above. If additional clarification is necessary, the possibility exists of providing a standard review plan or similar guidance to be more specific as to how and when these good practices should be implemented.

Other minor editorial comments about the scope of the report were also received and have been addressed in this version of the report.

Comment #2

The NRC received two comments referring to a possible lack of sufficient input from a broader set of stakeholders in the creation of the document. In particular, the comments cited the primarily NRC-related experience of the authors as being potentially too limited to produce such a document.

Response #2

As stated in the previous response, the initial motivation for this report was for use in internal NRC staff reviews of future HRAs, even though it is recognized this has implications about expectations regarding how HRAs should be performed. Hence, it was appropriate for persons most familiar with the needs and processes within the NRC to produce the report.

Nonetheless, and as further clarified in this report as a result of these comments, the authors collectively have both NRC and industry experience in HRA with at least two authors having worked on several industry PRAs including some recent licensee HRA submittals applied to specific issues. Based on this practical knowledge as well as review of available literature, the authors purposely attempted to cover the current practices of a variety of HRA techniques and tools commonly used in the nuclear industry.

Further, the NRC used the public review process and other formats to obtain input from external stakeholders including the nuclear industry, the Advisory Committee on Reactor Safeguards (ACRS), and most recently, international HRA practitioners.

Through the above processes, the report reflects a wide variety of inputs and views and, thus, is believed to represent a reasonable consensus as to typical good practices in HRA. In fact, the authors note that of all comments received, very few actually took issue with any of the good practices per se, and those comments tended to simply address particular points of clarification.

Comment #3

The NRC received comments suggesting that the good practices document should include guidance for improved use of existing data sources, both to help in identifying contextual factors that can influence performance and to support the quantification process. One suggestion was to look more carefully at existing operating experience data (several sources were proposed) for the specific plant, others of its type and any with common ownership, and for the industry as a whole, in order to identify factors that should be considered in assessing and quantifying human error probabilities (HEPs). That is, the data could be used to help identify actual causes of unsafe human actions in plants. In addition, the comments suggested that such data might provide insight about organizational factors that could influence the likelihood of unsafe human actions. Another suggestion was to use data from nuclear power plant simulator-based research, such as that conducted at the Halden Research Project in Norway, and operational event databases such as that being developed at the Idaho National Engineering and Environmental Laboratory and other international efforts, to support the quantification of HEPs.

With respect to using data to provide insights about organizational factors, in a related comment, the ACRS also suggested (on April 22, 2004) that guidance for the treatment of organizational factors in HRA would be very useful and should eventually be included in the good practices.

Response #3

The authors agree that the use of "empirical"data to support HRA is an excellent idea, and the collection and representation of such data are probably among the best ways to improve HRA in the future. In fact, the NRC is involved in several national and international efforts to develop databases to support HRA, including support of the Halden Research Project and the Human Event Repository and Analysis (HERA) database development at the Idaho National Engineering and Environmental Laboratory, as noted above. The NRC views such efforts as important contributors to improving HRA in the near future. However, adequate tools are not currently available to guide analysts in how to obtain and use such data to support an HRA, particularly in a quantitative manner. Many of the possible data sources were not developed for collecting information suitable to HRA, and the efforts to develop methods to optimize the usability of existing data sources have not yet reached a stage that is sufficiently complete and widely accessible. In other words, the state-of-the-art is not adequate to advocate such approaches as "good practices"; these suggestions, therefore, fall into the category of "best" or "better" practices. While the NRC encourages qualified HRA analysts to attempt to use data and develop processes for using such data, it seems premature to include such data and development activities as good practice for quantifying human error probabilities at this time. Nonetheless, the NRC does agree that the qualitative information from available data sources such as the types of errors being made and the conditions under which they are made can be helpful in more comprehensively identifying potential human errors of interest. Thus, recommendations to use available data in this manner have been added to this document. To the extent analysts attempt to use such data and take the steps necessary to incorporate the data appropriately, the HRA results should be more realistic.

A related argument can be made with respect to including guidance for treating organizational issues. While this report addresses performance-shaping factors that will help analysts to consider some organization-related influences, such as crew dynamics, characteristics, and potential biases and informal rules that may be operative (see Appendix B), the state-of-the-art in how to identify and understand important organizational influences and how to use that information in determining HEPs is not yet adequate. Thus, the NRC believes that it is not possible at this time to include an adequate set of good practices related specifically to organizational issues. Again, however, to the extent analysts attempt to address such issues, HRA practices will be improved.

Comment #4

The NRC received several comments regarding the inclusion of guidance for addressing errors of commission (EOCs) in the good practices document. Comments ranged from suggestions for how to better address EOCs (one of which was included in the document) to arguments that identifying and addressing EOCs is too resource-demanding for the anticipated gain, and actually not really necessary because EOCs are so unlikely.

Response #4

The latter argument does not seem appropriate as iterated in the discussion in Section 6 of this report. Techniques exist and continue to be improved, including work in the international community, for doing searches for EOCs — at least in narrowly defined areas that are not overly resource demanding. The push to include EOCs in PRAs is coming from the historical experience that serious nuclear power and industrial incidents have had EOCs as contributing factors and, at a minimum, future plant changes should be reviewed to ensure that the changes do not introduce conditions that are prone to or increase the likelihood of EOCs.

Comment #5

The NRC received comments questioning the appropriateness of screening out likely low-significant pre-initiator errors from the PRA model. The comments contained suggestions that it is better to maintain the error events in the model to capture their impact ,but perhaps simplify their quantification (e.g., use conservative values), and it is useful to model these events since most of the analysis of the errors (to decide they can be screened out) is performed anyway and future applications can benefit by having these events in the model. Additionally, a few comments thought certain screening criteria were inappropriate because they were either too restrictive or not restrictive enough.

Response #5

The good practices in this report are consistent with Regulatory Guide 1.200 and the ASME Standard, which both allow for screening pre-initiator activities "from further consideration." Nonetheless, the above comments have merit, and the authors have revised the report to reflect the potential benefits of not screening out these events from the model as an alternative approach. As for the specific criteria, they are very much consistent with those found under Capability Categories II and III in the supporting requirements of the ASME Standard with only slight modification to reflect current practices.

Comment #6

The remaining comments suggested a number of additions or clarifications regarding such topics as HRA team makeup, the goals of procedure reviews and what to review for, the post-initiator performance-shaping factors (PSFs), the usefulness of talkthroughs and simulator observations, dependency factors, recovery vs. repair, and other editorial suggestions. These comments tended to address specific statements in the document recommending changes to further improve the document.

Response #6

For the most part, the authors have revised the report to reflect these comments. In particular, the authors have incorporated most of the points into the report and, in some cases, provided further clarification to otherwise address the subject of the comment.

NRC FORM 335 (2-89) NRCM 1102, 3201, 3202	U.S. NUCLEAR REGULATORY COMMISSION BIBLIOGRAPHIC DATA SHEET (See instructions on the reverse)	1. REPORT NUMBER (Assigned by NRC, Add Vol., Supp., Rev., and Addendum Numbers, if any.) NUREG-1792

2. TITLE AND SUBTITLE

Good Practices for Implementing Human Reliability Analysis (HRA)

Final Report

3. DATE REPORT PUBLISHED	
MONTH	YEAR
April	2005

4. FIN OR GRANT NUMBER

5. AUTHOR(S)

Alan Kolaczkowski, Science Applications International Corp.
John Forester, Sandia National Laboratories
Erasmia Lois, U.S. Nuclear Regulatory Commission
Susan Cooper, U.S. Nuclear Regulatory Commission

6. TYPE OF REPORT

Technical

7. PERIOD COVERED *(Inclusive Dates)*

8. PERFORMING ORGANIZATION - NAME AND ADDRESS *(If NRC, provide Division, Office or Region, U.S. Nuclear Regulatory Commission, and mailing address; if contractor, provide name and mailing address.)*

Division of Risk Analysis and Applications
Office of Nuclear Regulatory Research
U.S. Nuclear Regulatory Commission
Washington, DC 20555-0001

Sandia National Laboratories
P.O. Box 5800, MS0748
Albuquerque, NM 87185

Science Applications International Corp.
405 Urban Street, Suite 400
Lakewood, CO 80220

9. SPONSORING ORGANIZATION - NAME AND ADDRESS *(If NRC, type "Same as above"; if contractor, provide NRC Division, Office or Region, U.S. Nuclear Regulatory Commission, and mailing address.)*

Same as above

10. SUPPLEMENTARY NOTES
E. Lois, NRC Project Manager

11. ABSTRACT *(200 words or less)*

The U.S. Nuclear Regulatory Commission is establishing "good practices" for performing human reliability analyses (HRAs) and reviewing HRAs to assess the quality of those analyses. The good practices were developed as part of the NRC's activities to address quality issues related to probabilistic risk assessment (PRA) and, as such, support the implementation of Regulatory Guide (RG) 1.200, "An Approach for Determining the Technical Adequacy of Probabilistic Risk Assessment Results for Risk-Informed Activities," dated February 2004.

The HRA good practices documented in this report are of a generic nature; that is, they are not tied to any specific methods or tools that could be employed to perform an HRA. As such, the good practices support the implementation of RG 1.200 for Level 1 and limited Level 2 internal event PRAs with the reactor at full power. Their elements are directly linked to RG 1.200, which reflects and endorses (with certain clarifications and substitutions) the "Standard for Probabilistic Risk Assessment for Nuclear Power Plant Applications" (RA-S-2002 and Addenda RA-Sa-2003) promulgated by the American Society of Mechanical Engineers, and "Probabilistic Risk Assessment (PRA) Peer Review Process Guidance" (NEI 00-02, Revision A3) promulgated by the Nuclear Energy Institute. This report is not intended to constitute a standard and, hence, it does not provide de facto requirements; rather, this report is intended for use as a reference guide. Consequently, the decisions regarding which good practices are applicable — and the extent to which those practices should be met — depends on the nature of the given regulatory application.

12. KEY WORDS/DESCRIPTORS *(List words or phrases that will assist researchers in locating the report.)*

human reliability analysis
human performance
HRA
probabilistic risk assessment
PRA
good practices
HRA guidance
HRA review guidance

13. AVAILABILITY STATEMENT

unlimited

14. SECURITY CLASSIFICATION

(This Page)
unclassified

(This Report)
unclassified

15. NUMBER OF PAGES

16. PRICE

NRC FORM 335 (2-89)

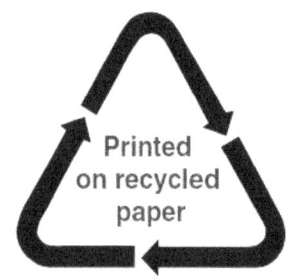

Printed on recycled paper

Federal Recycling Program